It's Not About the Tech

It's Not About the Tech

From Potholes to Piglets: Real-Life Lessons in Health IT

CT Lin, MD, FACP, FAMIA

ISBN 979-8-9969233-1-1
Library of Congress Control Number (LCCN): 2026917488

Editing: Brendan G. Craine
Typesetting & Cover: Paolo Biagini

The text of this book was written and edited wholly by humans, without the use of generative AI or LLM-based tools.

CT Lin is an excellent communicator.

—Jill VanWyk, MD

In It's Not About the Tech, *CT Lin brilliantly bridges the gap between digital innovation and clinical execution, proving that the true power of health information technology lies beyond the code itself. This book provides an essential roadmap for turning sophisticated tools into meaningful impacts for the patients and communities we serve.*

—Dr. Philip Payne, MD (President and Chair, American Medical Informatics Association)

Most health systems invest as if technology is the answer. This book politely (and repeatedly) proves otherwise. Through memorable anecdotes, it makes the case that transformation lives or dies in the space between workflow, incentives, and trust.

—Craig Joseph, MD (Chief Medical Officer, Nordic Global)

This book will undoubtedly guide the next generation as we build teams equipped with the skills vital for successful digital transformation. Dr. Lin's leadership, marked by grace and humility, will continue to light the path for those who follow.

—Margaret Lozovatsky, (Chief Digital and Information Officer, PremierHealth)

CT Lin spent thirty years learning what every health system relearns the hard way: The technology was never the hard part. Every CMIO knows the project that worked on paper and died at the bedside. CT names why.

—Adam Carewe, MD (Founder of NerdMDs and "Retired" CMIO)

An excellent treatise on the theory and practice of clinical informatics as taught by a master of the trade. This book is essential reading for all clinical informatics practitioners.

—William Hersh, MD (Professor at Division of Informatics, Clinical Epidemiology, and Translational Data Science, Oregon Health & Science University)

More from Dr. CT Lin at https://ctlin.blog/

To my parents,
please keep playing tennis into your 90s. And thanks for the genes.

To my kids,
whose unexpected wisdom constantly delights me.

To my wife,
thank you for your love, and unflagging support of my crazy ideas.

CONTENTS

ACKNOWLEDGMENTS

The ideas in this book owe their provenance to more people than I can name.

Thanks in particular to:

Steve Hess, who gave me the latitude and encouragement to create and grow informatics from the ground up.

My Senior Medical Directors—Jonathan Pell, Amber Sieja, Heather Holmstrom, Kelly Bookman, Katie Markley, Jennifer Simpson.

My Mentee Wisdom Circle—Esther Hyelim Park, Jennifer Goldman, Matthew Tien, Christina Jung, Maruf Haider.

My readers, who helped make my chapters better—Betsy Toll, Melissa Jost, Jill VanWyk, Steve Rotholz, Amy Hassell, John Borleske, Jashanpreet Singh, Mark Earnest, Liz Salmi, Eric Glissmeyer, Steve Ross, Margaret Lozovatsky, Eric Helsher, Steve Clark, Jeffrey Koepp, Jeff Sippel, Peg Reidy, Eden English, Richard Altman, Kate Perica, Katy Trinkley, Jennifer Simpson, Vineet Chopra, Tyler Anstett, Ann Boyer, Bryan Steitz, Lucinda Welch, Cindy Larson, Kate Sanfilippo, Alice Pekarek, Sam Butler, Steven Nordstrom, Ed Horowitz, Aram Alexanian, Laura Macke, Sigall Bell, Fabienne Bourgeois, Marshall Davis, Foster Goss.

My colleagues, too many to name, for everything.

And my editor, Brendan Craine—your cortex shaped a jumble of chapters into a cohesive journey and messy thoughts into sharper insights. May we take more journeys together.

ACKNOWLEDGMENTS

[illegible]

INTRODUCTION

Sepsis and the Angel in the Sky

Sepsis is a hidden killer. The infection sneaks into the bloodstream from pneumonia, kidney infection, or even bone fractures. Every hour, 50 people die from sepsis in the United States, and hospitalizations for sepsis have doubled in the past 10 years.

What can we do about this?

Early-stage sepsis is tough to spot: uneven blood pressures, uneven body temperature, a slight increase in heart rate or breathing. These changes are easily hidden within the normal pattern of vital signs. And yet, the earlier sepsis is found and treated, the better a patient's chance of survival.

In 2018, my colleague Amy Hassell, Chief Nursing Officer of the Virtual Health Center (VHC) at UCHealth, was the driving force behind our push to defeat sepsis. She knew too well how many patients were lost to bloodstream infection. She knew that AI predictive models might spot a patient with sepsis faster than a human could. But could AI models work in a real hospital?

With the help of a data scientist, she assembled four predictive AI models whose outputs generated a risk score. The system would monitor patients' vitals and send an alert if someone was at high risk of developing sepsis in the next 12 hours. The sensitivity of the ensemble model was 96% and the specificity was 88%. Mathematically, this was excellent performance. The big question was, would it improve care at the bedside?

With great fanfare, we launched our sepsis predictive AI model in six medical and surgical units, letting it watch over hundreds of patients. The doctors and nurses on these units could see the color-coded scores for every patient in real-time.

So, what happened?

After three months of displaying the score: nothing. No change in number of deaths from sepsis or transfers to ICU for patients getting sicker. We were not faster at giving extra IV fluids or antibiotics. The AI models seemed completely ineffective.

The first hint of trouble was the number of alerts per day. Each hospital unit of about 30 beds received 61 alerts each day, of which only *two* were true-positive for sepsis. This is a 30-to-1 ratio, or just over three percent. Too much noise for too little signal.

Imagine being a frontline nurse. The system pings: "Alert on patient in bed five." You go read the chart, check on the patient, and after a few minutes of study, decide it's not really sepsis. Another ping: "Alert on bed seven." Check the patient. Again, not sepsis. "Alert on bed 14." Check the patient. Not sepsis. Meanwhile, your regular work is piling up. Instead of caring for patients who are sick right now, our predictive AI is sending you to patients who might get sick in 12 hours. Even worse, when you explain your concern to our predictive model team, they tell you "Don't worry, keep going! Too bad about the false alarms, but if you check every alert, you *will* find every sepsis case. Trust us!"

You would walk away in disgust. "I don't have time for this."

Both of you are right: The math detected almost every sepsis case. But the bedside clinical team was too busy taking care of patients who were sick *right then*. No one has extra time to find the needle in the haystack, the patient with early sepsis, who will become deathly ill in 12 hours.

Frustrated, we turned off the predictive alerts. Was this project doomed? We went back to the team room and mulled it over. We believed in the paradigm shift that could come from predictive AI. We had a tool that could glimpse the future. What we apparently didn't have was a team or a process that could use the tool effectively.

Then came CNO Amy's big insight: She would gather a handful of dedicated ICU nurses and place them in a dark room in a remote building with a bank of widescreen monitors. (That building would later become the Virtual Health Center). Protected from distractions, this team would review all the alerts, quickly assess the patient's EHR data, check the in-room remote camera, and identify the true-positive early sepsis patients.

Weeks later, with this new team, we launched our second AI trial. We pulled the alert system out of the hospital units and only showed alerts to the VHC nurses. They would dive into the haystack and find the needle—the highest-risk patient—and contact the bedside nurse and doctor to start sepsis treatment.

The result? It worked. Not only did we detect sepsis earlier, we were able to administer IV fluids and antibiotics more quickly. The time from sepsis detection to starting IV fluids fell from 1 hour and 36 minutes to a mere 19 minutes, and time to starting antibiotics went from 1 hour and 34 minutes to less than half that. Over the following year, we saved 211 more lives from sepsis.

A smashing success! Right?

Alas, no. All was not well around the fields and haystacks. Our VHC nurses reported unexpected resistance. Often, when the VHC nurse called the bedside nurse or doctor, they were met with incredulity. "Why are you calling me about *my* patient? They look fine. I don't agree that they have sepsis. I don't agree with calling the Rapid Response Team." In the worst cases, our bedside nurses hung jackets over the in-room cameras. In effect, they were telling us, "We don't trust you."

How do you dispel distrust? It took a road show. We sent our VHC nurses to meet the bedside nurses in person. In turn, the bedside nurses were surprised when the disembodied voices from the VHC belonged to ICU nurses with decades of bedside experience. The alerts, they realized, weren't coming from faceless voices; they were coming from peers. Even then, the trust grew slowly, and the culture shift took months.

Thanks to that carefully rebuilt relationship, our bedside nurses now introduce the VHC nurse to the patient as "our angel in the sky," and say that "they help me keep you safe." The AI prediction now includes other deteriorations: alcohol withdrawal, pneumonia, heart attack. With further improvements, we extended the ability of one VHC nurse from monitoring 150 patients simultaneously to monitoring 500. In 2025, mortality from in-hospital deterioration improved so much that we saved 1,000 more lives per year across 14 hospitals compared to our baseline.

Technology alone was insufficient to improve sepsis care. As we will see, technology itself determines only about 20% of the success of a technology project. Even when the predictive AI tool worked, it didn't save lives. We had to reconfigure our teams and rebuild trust. To properly implement the technology, we had to address the human side of the problem.

The field of Health IT (Information Technology) is about the flow of information and the technologies deployed in healthcare. The field of informatics is about the delicate interplay between clinical medicine and these healthcare tools. Informatics is the common thread through this book, and it has everything to do with success or failure of Health IT projects.

Those who work outside informatics think, "Cool technology. Plug it in and it works, right?" But nothing could be further from the truth.

What Is "the Psycho 80"?

Early in my career, in 1999, I traveled to Salt Lake City to visit the headquarters of 3M Health Information Systems. While there, I took a side trip to visit Reed Gardner, PhD, an informaticist at LDS Hospital and one of the founders of what became medical informatics. One of the many things he taught me was that information technology projects are a triad: **people**, **process**, and **technology**. Too often we focus on the technology (that expensive thing we bought) but pay little attention to the people (Who will use it? Why? Do they need training?) and the process (How and when will they use it? Will it be efficient?).

Furthermore, projects are not equal parts people, process, and technology. It is difficult to get the **technology** right, of course. But when projects fail, **people** and **process** are usually the reasons. Only 20% of a project's success is due to the technology; 80% of the success is due to the psycho-socio-political skills of the people doing the work and the process that is crafted as a result. In other words, it is *four times harder* to manage humans than to wrangle technology.

"Surely," you say, "informaticists must already know this?"

No, we get this wrong all the time. Many good ideas have failed and still fail because they were not implemented in a way that accounted for people and process.

Here's a personal example: I built an alternative template for progress notes that was easier to read. I was confident that it would be widely adopted, but I didn't explain the reasons sufficiently. Very few colleagues used it. I had a **people** problem. (see **APSO Notes, Maxwell, and the Meeting Before the Meeting** on page 98).

Another example: We built visual pathways in the EHR that simplified complicated diseases and made it easier to diagnose and treat them. The pathways were only used sporadically, since doctors had to pause and click to a different screen to use them. We had a **process** problem.

People and process pitfalls are the traps that foil your project and send it down the trash chute, into the dumpster of history. They are crucial factors in your project's success. I call them the Psycho 80.

Too often, we get tunnel vision looking at the shiny technology and don't widen our vision to include people and process. If my life can serve as a warning to others, then I would plant a bright red "Psycho 80" warning flag in front of all the potholes I have stepped in over the years.

How might we sidestep these pitfalls and potholes? Much of this book is dedicated to stories of stepping in those potholes, and out of them, and the lessons I learned.

I believe that physician informaticists are critical to the future of healthcare. Populations are aging. Members of our care teams can now be far apart and virtually connected. Patient data grows exponentially. Knowledge of human disease grows exponentially. Technology accelerates exponentially. The ground is constantly shifting under our feet; the best-practice patient care changes every day.

Informaticists who understand clinical care and technology and change management are positioned to bridge these chaotic realms and shepherd in the next generation of healthcare. But their work must encompass more than technology. A failure to address the Psycho 80 spells doom for any informatics project.

Earlier Is Better

We like to think that we run a mature informatics program. And yet, we still sometimes ask ourselves, "How did this clinical decision support tool get built without any of us hearing about it?"

A few years ago, primary care doctors at our organization had a tough time ordering the right cardiac monitor. It was hard to order the right test for each patient. Which test to order for heart palpitations, or dizziness? What about a follow-up on known cardiac arrhythmia? How to choose between these new ZIO Patch monitors and older Holter monitors? These decisions required wide-ranging considerations, which the doctors had to navigate themselves.

One of our electrophysiology cardiologists took up the cause and contacted an EHR analyst directly. The pair spent several months designing a new cardiac monitoring order, including detailed descriptions and lots of branching logic: Does your clinic stock this device? Do you want the company to mail it directly to the patient? Do you need "live monitoring" because of imminent risk? Do you want a recording duration of 48 hours? Two weeks? A month? What kind of arrhythmia are you expecting? Each one of these questions included paragraphs of explanation to guide the decision.

This comprehensive but complicated order was implemented. Immediately, our primary care physician (PCP) colleagues called us to complain: Dense paragraphs of reading made the order hard to use. PCPs still ordered the wrong tests, and the cardiology clinic had a spike in calls from confused docs. No one was happy.

Without an informaticist, an organization could have spent weeks or even months working this out. The cardiology expert would insist on a comprehensively correct order. The PCPs would want a simple order. The IT analyst would be caught in the middle, trying to choose between many EHR design possibilities.

In the uproar, one of our senior informaticists jumped in. Gathering feedback from a few PCPs and applying her own internal medicine knowledge, she met with the cardiologist to work out a simpler order. What if we reduced the options and made it easy to order the most common tests? We could strip out 80% of the verbiage and guide the PCPs to one of three clear choices.

In a one-hour meeting, they clarified the goals and simplified the order. It was straightforward to rebuild and was in use by the next week. PCPs were much happier, and cardiac monitor orders started to flow smoothly.

This was a quick and successful emergency repair. Hooray!

This is where informatics shines: We speak the language of clinical medicine, we get what EHRs can do, and we know how to "make the right thing easy."

But could we have avoided this emergency in the first place?

Unless informaticists are present at the design phase of a project, we will be forced to come in later for emergency repairs. Or worse yet, we will never become involved, leading to poorly designed EHR tools that are never repaired and continually frustrate physician users.

What Clinical Informatics Is NOT

I recently read a book called *Mindfulness in Plain English*[1]. I loved how the author described meditation in terms of what it is NOT.

WHAT MEDITATION IS NOT (misconceptions)

- Meditation is just relaxation
- Meditation is mysterious and cannot be understood
- Meditation is for saints, not regular people
- Meditation is running away from reality

Great idea. Let's write this for informatics.

WHAT CLINICAL INFORMATICS IS NOT (misconceptions)

- Informatics helps set up your computer (no, that's Information Technology)
- Informatics is quick and easy (no, it's never "just put a hard stop and done!")
- Informatics people like computers more than talking to humans (please, no)
- Informatics forces people to do things they don't want to do (no, no, no)

Our physician informatics group has a logo. It shows a PIG (member of our Physician Informatics Group—no, we don't take ourselves too seriously) at a computer next to a puzzle, a handshake, and some arrows. Our tagline: "From Complexity, through Relationships, to Solutions."

Physician/APP **Informatics** Group
From Complexity, through Relationships, to Solutions

In recent years, many health systems have installed electronic health records. But without clinical informaticists, all you have is an expensive ship and no captain. Informaticists help steer the ship so that doctors and patients get where they want to go.

A note on terminology:
Physicians, **Advanced Practice Professionals**, **Clinicians**, and **Providers**. Many consider these terms interchangeable. In the beginning of my career, I had no experience with APPs (nurse practitioners, physician assistants, midwives). For simplicity, many use the term "provider," as in "Health Care Provider," to indicate physicians and APPs. Others use the term "clinician," which some say includes nurses and other clinical assistants. I mainly use "physician" and sometimes "physician/APP" in my writing, as this combined population has been the focus of my work.

Medical Informatics, **Clinical Informatics**, **Physician Informatics** and **Health IT**. I consider medical and physician informatics to be interchangeable in this book. Clinical Informatics is broader, including physicians/APPs as well as nurses, pharmacists, and other healthcare professionals. Health IT is the yet broader scope of all information technologies used in healthcare, whereas medical informatics is the intersection between physician/APP expertise, deep knowledge of clinical workflows, and application of Health IT.

The Four "Ins" of Informatics

This book recounts moments of despair and joy, told through stories and ideas. I sort them into four major tasks of practical informatics. Steal these ideas and apply them for yourself. Like INformatics, they all happen to start with IN: INvestigation, INtrigue, INfluence, and INosculation (yes, no one knows what this means).

INVESTIGATE: How complex is the status quo? What are the differences between clinical medicine and information technology? Medicine and IT are less like peanut butter and jelly and more like oil and water. What are the consequences when these two fields do not get along?

INTRIGUE: How might we get clinicians interested in IT? How might we teach IT colleagues that the pressures of clinical medicine will eat technology for breakfast? We must hone our skills to capture attention and spark insight.

INOSCULATE: "To unite by fitting closely into each other." Originally meant to describe blood vessel anatomy, it also applies to medicine and IT. We must build trust, build consensus, build relationships. Healthcare technology does not work unless used well by clinicians and patients.

INFLUENCE: Lead the change. We make small changes and nudge individual behavior. We make large changes and shift cultural norms to improve the lives of patients and clinicians. How might we decide when to go big?

The final two sections answer broad but important questions:

How can you THRIVE personally and professionally? Here are the habits, skills and ideas I use in my daily work. I hope some of them are beneficial to you.

What will THE FUTURE hold? Medicine is changing quickly and becoming more complex. Informaticists already deal with clinician burnout, reductions in the workforce, and disruptive technologies like generative AI. What is coming next?

How to Read This Book

If you are...
A **junior clinical informaticist**:
Read the PIG-let questions and see where they take you. Look up chapters that interest you. You are my main audience; clinical informaticists are the future of our profession.

One of my **mentees**:
Some of your stories and lessons are here. I have learned a lot from you, and I am grateful. Are there any ideas we have not yet discussed? We can, at our next meeting.

A **new PIGlet** (member of my Physician/APP Informatics Group):
This book is required reading and a way to understand how your CMIO thinks. This is our informatics journey. Avoid the potholes I stepped in. Make new mistakes, not old ones.

A **senior informaticist**:
You will recognize some dragons we slayed together. Thank you for being my colleague. You'll chuckle at my failures. Will an idea I mention help your team?

A **student, resident, or junior faculty**, wondering "What is informatics?":
Read the Table of Contents, Intro, and Conclusion. I hope one of the stories sparks an insight. Learn why I believe informatics is a key to the future of medicine.

A **leader in healthcare, or any field that relies on people, process and technology**:
See if the skills and strategies resonate with you. We work in complex environments. Often there are no rules. We learn from trial and error and

from each other.

A **member of my family**:
I am sorry for those long hours hunched over a desk and being grumpy. Thank you for listening to my stories, sometimes more than once! Thank you for your love and support.

My Handoff to You

This is the book I wish I'd had 30 years ago. I could have used these ideas and skills to avoid many of my failures. Those failures, however, were the lessons that fueled my later success. In writing this book, I hope to identify my mistakes and help you avoid them.

Explaining problems and solutions with theory is *boring*. Too often, it becomes a series of self-evident statements, like "The general principle is that we love Mom and apple pie." Of course we love Mom and apple pie. Duh. Everyone loves Mom and apple pie. The statement is so bland it has no meaning. Instead, we need a story. "Last week, Mom made us an apple pie, but something wasn't right. This one time..." etc.

I tell *specific stories* so they are more likely to stick with you. However, because they are my experiences, they are unavoidably egocentric. For that, I apologize. (And include stories of failure for comic relief.)

Technology is changing and accelerating. What chance do I have to teach you something useful about the unknown future? Past examples may not help. Instead, we focus on human behavior, which does not change quickly. Psycho 80 skills will serve you well for many years.

My stories are intended as stepping-stones. You, the thoughtful and forward-thinking informaticist, will be architect and engineer, solving thorny problems we have not yet faced with tools we have never imagined. I can't wait to see the future you will build.

— CT Lin MD
Professor, University of Colorado School of Medicine
Chief Medical Information Officer, UCHealth—Colorado (2006-2026)
Fellow, American College of Physicians
Fellow, American Medical Informatics Association
Bad ukulele, Good medicine
Denver, CO, May, 2026

17 PIGlet Questions

One of the perks of being a leader is choosing names for my committees. I like to pick ones that will inject a moment of joy into my work. For example, I enjoy participating in...

- JIG (Joint Informatics Group). During the pandemic, this became the "Daily JIG."
- EMO (Electronic Medication order Optimization).
- WONK (Workgroup Orderset Nowledge Kommittee). No one said it had to be spelled right, and it is lovely to hear colleagues say, "Take that proposal to WONK."
- PIG (Physician/APP Informatics Group). The big meeting with my entire informatics team is the "Large PIG." The Senior PIG is my senior medical directors, although "PIG heads" might be more accurate.
- I even tried to name a national committee Group of Users at Academic centers, National Organization (GUANO). Tried, and failed. But you don't succeed if you don't try.

My newest informatics recruit asked, "If Large PIG is our main group, does that make me a PIGlet?" Yes, I think it does. In support of our PIGlets, here are 17 questions frequently asked by PIGlets, junior informaticists, and my mentees, with references to the chapters that discuss them.

Which questions and chapters apply to you?

Question	Chapter
What is informatics, anyway?	INTRODUCTION
I want to do a project, but how do I get started?	Minimum Viable Product (INFLUENCE)

Question	Chapter
I have an idea. How can I get help?	The Superpower That Is the One-Pager (INTRIGUE)
How can I avoid doing this assigned project?	Don't Be Selfish, Be Useful (INFLUENCE)
Where do I find more time?	Getting Things Done in an Age of Acceleration (THRIVE)
Why does no one agree with my great idea?	What Is Your Boss's Learning Style? (INTRIGUE))
They treat me like a help desk. What do I do?	Subterranean Informatics (INFLUENCE)
They bring me in too late. What do I do?	Earlier Is Better (INTRODUCTION)
How do I write my resume?	My Failure Resume (THRIVE)
How do I publish a paper?	Write Like Hell (THRIVE)
How can I get a title so that I'm recognized?	You Don't Want a Title; What You Want Is Time (THRIVE)
How do I give a good presentation?	The DOs and DON'Ts of an Effective Presentation (INTRIGUE)
How do I choose my next project?	Three-Circle Venn Diagram (INFLUENCE)
I have a team! What now?	What Is Your Vision Statement? (INFLUENCE)
I'm surviving, but how can I thrive?	THRIVE

Question	Chapter
How can I make my job joyful?	Get in "The Zone" (THRIVE)
When should I go all-in?	Punctuated Equilibrium (THRIVE)

INVESTIGATION

Informatics is about listening hard and listening well. One must understand the pressures of being a clinician and having patient lives on the line. One must also understand what technology can and cannot do. Your work should begin by investigating the complex ways these two fields interact.

Intro—The Frog at the Bottom of the Well

There is a Chinese proverb: *jǐng dǐ zhī wā*. "The frog at the bottom of the well." Or, in full, "the frog at the bottom of the well believes the sky is just a tiny blue disk."

Like the frog in the well, some people sit where they see only a limited view of the world. Physicians practice in one specialty, in one clinic, or focus on one service. They are very good at their work, and a focused view is appropriate.

On the other hand, informaticists design an EHR to be used by everyone. Changes made in these "commons" affect all users. Informatics frogs sit in a meadow, look at the big blue sky, and realize there are hundreds of wells with frogs at the bottom of all of them, requesting improvements to the EHR.

Example 1. A cardiologist believes that "minutes exercised per week" is an important vital sign of a patient's heart health. He wants this data to be captured at every primary care clinic visit. In fact, he believes that number is so important that it should be a "hard stop" question required at every primary care appointment.

The frog at the bottom of the well thinks to itself, "How hard can this be? My idea is simple. The EHR team can add a mandatory question like 'How many minutes do you exercise every week?' Then the medical assistant will record it at every visit." The cardiologist frog thinks, "This is good for patient care. The medical profession should pay more attention to exercise, and tracking this number is the first step." The frog also thinks, "This will help me in my research project."

Meanwhile, informatics frogs see the big blue sky and clouds on the horizon. Calling attention to regular exercise is a good idea—in theory. But informatics frogs also see the hundreds of other wells with hundreds of other frogs, each with their own priorities and demands. In our organization, doctors see several million patient visits per year, and this mandatory question will add two minutes to every visit. In pleasing one frog, we would encounter a full-throated chorus of unhappy frogs. As a rain cloud might replenish one dry well, other wells might overflow.

Our informaticists handled this exact scenario. We replied to the cardiologist, "If you can get the leaders of primary care clinics to agree with your idea, we would be happy to make this change."

We did not hear back from that cardiologist again.

Example 2. In a previous EHR, we managed a "scanned documents" folder. We received many paper documents and allowed requesters to name new document types as they preferred. At first, it worked well. Our team quickly agreed with each request. Each bottom-of-the-well frog got their way.

Soon, however, we had over 500 scan categories, and no one could find anything in the scanned documents folder. Similar documents had different names and were sorted alphabetically: Advance Directives. Delegated Decision-Maker. Five Wishes. MOST form. Patient Directives. These could be hidden across a dozen or more folders. The sky wasn't blue anymore. It rained on all of us.

When it started to rain, our informaticists took charge of the scan folder, gathered consensus, collapsed the categories, and renamed the documents to make them easier to find as "advance directives." Informatics frogs to the rescue.

Bottom-of-the-well frogs with parochial attitudes say, "That bureaucracy slows me down. If it were not for them, I could get things done." The informatics frog says, "Thanks for your request. Let's build something that works for everyone." When we can, we make sure everyone wins. With the tools we have, sometimes we have to say "no."

A key task of the informaticist is to balance local customization (making doctors happy and allowing different practice styles) with eliminating unnecessary variation (standardized work leads to efficiency).

How do informaticists learn to do this well? Deep investigation. Practice, thoughtful reflection, and experience. One must kiss a lot of frogs to discover your informatics prince. But that is mixing Disney fables and Chinese proverbs.

The Lesson Learned

Balance the parochial with the wider view. The passionate frontline clinician has a good local idea. Excessive variation is error-prone and inefficient. Both are important. Raise the frog from the bottom of the well so that their viewpoint changes and you work together. As we will explore, this is the core idea of our section on "Investigation." Understanding how clinicians think, and what came before, and how the larger system fits together, is crucial to the difficult task of introducing a new technology.

Stat-LAN and the Chief Complainer

Kids have ears and no mouth.

—Taiwanese proverb

Flashback—1997. I was a new assistant professor at the University of Colorado. We were a small campus of 200 doctors, two dozen clinics, and a 300-bed academic hospital. The patient charts where we wrote our notes and tracked test results were delivered every day on carts. Each clinic had recently installed a "dumb terminal" to run Stat-LAN, a laboratory results computer. Some doctors had learned to dictate their notes by telephone. In other words, electronic tools in our clinics were still a new thing, and they did not work very well.

One day, I received a printed memo in a brown interoffice envelope from our Chief Operating Officer (COO). The memo told us that physicians should be grateful for the new tools the hospital leaders had provided. "For example, we pay for the transcription service when you dictate your patient reports. Just talk into the telephone, and someone will type it for you."

"Also," the memo continued, "we bought a laboratory computer system to show you all the lab results you ordered on patients. We store *three months* of data, because really, what more do you need to take care of patients? And everything is really user-friendly. The login is 'lab,' and the password is 'lab'! That's as easy as we could make it."

I read through the memo with skepticism. This was ridiculous. After gathering feedback from my colleagues, I sat at the clinic's one IBM Selectric typewriter and typed out a prickly reply to the COO—a multi-page, point-by-point rebuttal to that widely distributed interoffice memo. I wrote many complaints about Stat-LAN, the new lab results system.

Yes, this new instant access to test results on a computer is great, but setting both the login and password to "lab" is a terrible security decision.

Plus, the Stat-LAN lab system is set up poorly. It can only display results in a long-scrolling reverse chronological order. To see labs performed two months ago, I have to press "page down" 10 or 15 times to find the "TSH" thyroid result that I was looking for. There is no search or index.

Also, storing only three months of lab results is shortsighted. No doctor can remember how long ago a particular lab was drawn. After searching the lab computer and finding nothing, one must also request the paper chart from archive, which takes another day or so.

Why is there is no digital "inbox" where I can receive the results I ordered? I have to search for each result individually. The typical primary care doc is responsible for over 1,000 patients. If I was worried about a particular result, I would bring the patient back for an appointment, or hope that the lab would fax it to me. Stat-LAN does not reduce my work.

Who bought this system? No doctor was consulted. We won't use it!

After skewering Stat-LAN, I also wrote my complaints about the dictation system.

Accessing the dictation service was clunky. A single dictation session involved dialing 30 numbers on the landline phone. (Remember, this was before cellphones and certainly before smartphones.) Press 9 to get out of local network, punch in the 10-digit phone number for the dictation service, then punch in a 6-digit day of service, 6-digit personal dictation code, and then the 7-digit medical record number so that the finished transcription could be delivered to the right doctor for a pen-and-ink signature. Imagine how often we mistyped one of those 30 numbers for dozens of dictations per day. Was it any wonder that dictations were frequently lost, or got filed in the wrong chart?

By way of summary, I proclaimed that these were Stupid Decisions and sent my scathing memo on its way.

A few weeks after my memo was circulated among hospital leaders, I was offered a job by the Chief Information Officer (CIO). There was no precedent for a doctor working in the Information Technology department, so he called my new job "Physician Liaison to the IT department." More accurately, I was Chief Complainer. He offered me 0.1 FTE (full-time equivalent). This meant the hospital would pay for four hours a week of my time, excuse me from seeing patients and ask me to attend two or three committee meetings a month.

I was very excited. I was pretty sure my job would involve telling computer programmers where to put the buttons on the screen and how to design the next computer module to help doctors do their work. I had just read Donald Norman's *The Design of Everyday Things*[1], and I was ready to apply design principles to computers in medicine. I took the job.

As an excited newbie Physician Liaison, I knew that the IT department was ineffective and did not understand doctors. They weren't working on what was important! In meetings, I would stomp my feet to make my points!

Looking back years later, my IT colleagues would tell me that my newcomer attitude in 1997 was "adorable."

In the first months of meeting with the CIO and his leaders, I learned how wrong I had been. For example, I had made fun of the initial investment in a lab system that would only keep three months of results. IT leaders were already planning to expand storage at a pace that the organization could afford. The 90-day time limit was temporary.

Oh.

I had also complained about how frequently our computer systems went down. I did *not* know that IT already had plans to build redundancy into the network, the servers, and the T1 lines that connected the outlying clinics. When I attended those planning meetings, for every concern I brought up, there were plans to make it better. Hmm. These people were pretty smart!

I felt silly. I had made assumptions. Had I been more curious, or listened more carefully to the technical reports, I could have avoided some misunderstandings.

On the other hand, I could see an important role for myself: translator and communicator. Our technologists were investing in the right tools, but no one was hearing about them. I started a monthly newsletter to update my clinical colleagues about the computers and the IT improvements that were coming. Soon, my colleagues began telling me, "Thanks for making such a difference over there." And all I had done was explain what was already in progress.

The Lesson Learned

Start with curiosity. It is surprising how often we don't. Assume your "opponent" or colleague or patient is thoughtful. In any healthcare organization, no one knows everything. Our complex work can easily lead to conflict. Any attempt to make improvements must start with clear understanding on all sides. In 1997, I learned that seeking this understanding was my job. This is STILL my job. Even now, I often fail to seek understanding first.

There is a Taiwanese saying: "Kids have ears and no mouth." When I first heard this at age 10, it made no sense to me. What do you mean? I have a mouth right here! My father gave me a long silent look and did not explain.

Decades later, I think he meant that kids need to listen, not talk, because you don't know anything yet. My father would have been a good informaticist. Had I been paying attention, he was teaching me to listen, the first step when we INVESTIGATE.

How It All Started

Time for the wayback machine. I get to put on my starch-stiffened authority-figure white coat as I tell you, "All y'all EHR-using folks don't know how good you have it." Today, looking up a medical record takes only a few seconds: *Tap-tap-tap*. This is infinitely better than what it used to be: "Call medical records and have them send a courier. I need that paper chart immediately!" (Which probably means tomorrow.)

Analog Photo #1: Medical Charts Stacked and Locked

For fun, I've set my Zoom background to a 1997 photo of medical records in the basement of University of Colorado Hospital. Each aisle stored thousands of charts, with more charts piled on top. The 29 aisles in the Record Room stored the last three years of charts; older records sat for 27 more years in a downtown warehouse.

Fun fact: We turned down many innovation partnerships and offers of free services because the medical information locked in those paper records was too difficult to retrieve. Imagine our disappointment when we had to decline an offer like, "We have a Pulmonary Function mobile van parked out front: send us all your patients who currently smoke, and we'll screen their lung function for free!" Or, "Our insurance company will send a bonus payment if you can prove that 70% of your patients who have had a heart attack are taking aspirin!" True story—a clinic using paper records ended up paying more in overtime wages to medical records staff gathering the data than the clinic later received from the quality bonus payment.

The difference between digital and physical record storage is occasionally stark: "Uh-oh! The medication Bextra is being recalled by the manufacturer! Let's tell our patients to stop taking that medication." Half our clinics ran a report and called patients immediately. The other half, who were still using paper records, had to wait for their patients to call for a refill a few months later, THEN tell them.

Analog Photo #2: Medical Records Intake Room

Our medical record staff filed an average of six vertical feet of paper every day. At the time, we were one hospital and 40 clinics. We employed 50 people and were constantly two weeks behind. If you asked for a chart, you couldn't expect to find anything in it from the last two weeks.

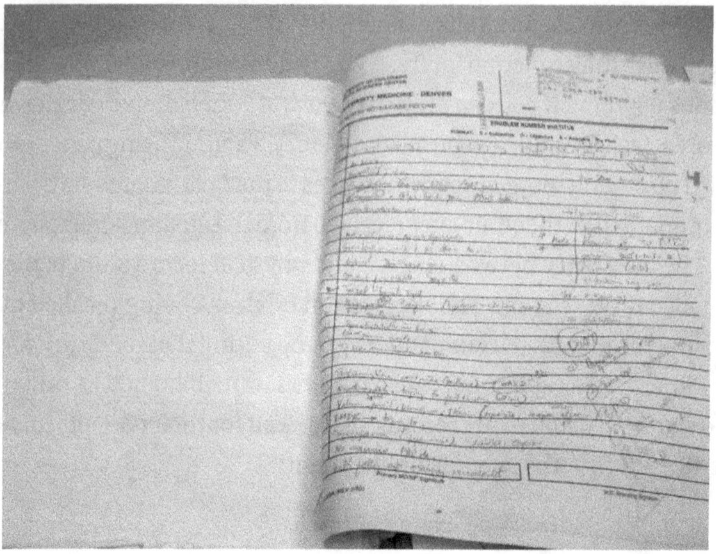

Analog Photo #3: Shadow Charts and Dictation

This is a paper progress note I scribbled in 1999 on non-carbon paper for a clinic patient. We sent the original to Hospital Medical Records and then filed the yellow copy in a "shadow chart." A shadow chart is a duplicate kept at our clinic. We didn't trust Hospital Medical Records to consistently deliver the charts for patients scheduled each day. The result? In addition to the 50 medical records staff at Hospital Medical Records, we had one or two additional staff at each of 40 clinics to keep the shadow charts updated.

The careful observer in Analog Photo #3 will see that I scribbled "[DICT]," meaning that I also dictated this note. This ensured there would be a more legible version available a few weeks later. Every evening, our clinic sent Dictaphone microcassette tapes to our transcription service, transported by human courier. Some of us remember hearing doctors fast-mumbling dictations while they walked the halls to their next patients. On average, transcriptions took about two weeks to type, mail, and file into the record. But hospital patients needed transcriptions returned immediately, so these notes were ordered STAT for three times the cost, typed the same day, arriving by urgent courier overnight and taped into the paper chart.

What if there was an urgent phone call, and we needed a particular medical record? We would routinely have staff members checking our shadow chart room, then wandering from desk to desk, asking: “Do you have the chart for Peterson, Mary, or Smith, Joseph, or Samuels, Jane?” and thus not available to answer the phone, or rooming patients…

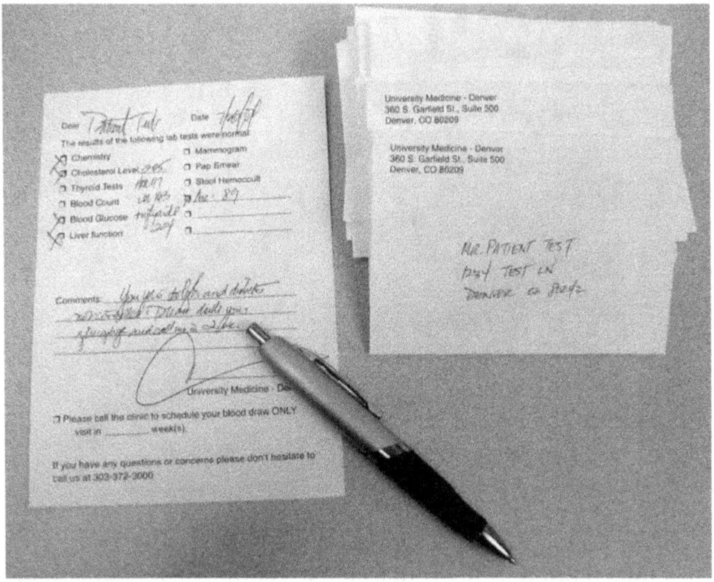

Analog Photo #4: Results by Postcard

When we ordered blood tests during patient visits, the results would come back a few days later by fax. “Peter Smith: Basic Metabolic Panel normal.” To notify patients, we would either ask our medical assistants to call and tell the patient (time consuming) or we would scribble something on a postcard. Every evening, I faced a stack of faxed reports and blank postcards. Please don’t ask me how many times a patient brought back a postcard, saying “This looks pretty important, but I think you meant to send this to a *different* Peter Smith.”

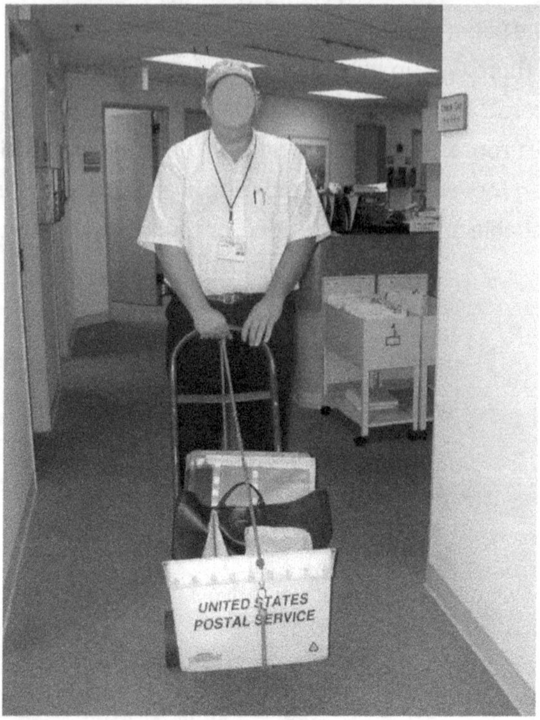

Analog Photo #5: Data in Rush-Hour Traffic

Our hardworking courier used a handcart to transport our precious medical information: X-ray films, paper records, Dictaphone tapes. His van and our data moved at the speed of rush-hour traffic on Colfax Avenue: seven miles round trip, twice a day, connecting 12 buildings around Denver.

How did we do with all this person-power? On a typical day, I would see 18 internal medicine patients. Main medical records would deliver nine charts, and our clinic delivered six shadow charts. This left, on average, *three* patients with no chart, just today's piece of paper, with their name, handwritten vitals, the "chief complaint," and stated allergies. Let me illustrate.

Imagine me standing outside the exam room, reading a single sheet of paper clipped next to the door:

Josie Smith
DOB: 1/2/1965
Chief Complaint: "Follow-up cardiology consult"
Vitals: BP 130/85, HR 72, Temp 98.6, Allergies: none

[What follows is an actual conversation.]

"Hi Doc! Great to see you! What did my cardiologist tell you about my visit two weeks ago? He said he sent you his report and that you'd explain."

"Um. I don't have any of your records today. But I see your blood pressure looks good and that you report no allergies to medicines."

"What? I made this appointment to go over his report! That visit was two weeks ago!"

"Yes. Um. What condition do you have? Why did we send you to cardiology? What do you remember him telling you?"

"This is disappointing. You mean you really have nothing? Do you at least have the blood tests or the echo result?"

"Um, no. I'm really sorry. I tell you what—no charge for today, I'm sorry to waste your time. I have requested your chart again, and I will call you later this week if your chart shows up after I yell at my medical records people."

"Whatever. You guys should really get your act together. Okay, can you at least refill those three medicines that you prescribed for me last year? I'm running out."

(I excitedly took out prescription pad.) "Sure, I'm happy to! Do you remember the names of the meds, and the doses, and what they're for...?"

Three-Ring Binders

Patients calling into our appointment scheduling system could expect a response along these lines: "Oh yes, thanks for calling! You're looking for Dr. Lin's next available appointment? Sorry, nothing for the next three weeks. Oh, you'd like to see the next available doctor? ... [quiet gnashing of teeth] Okay, I'll pull down the other twelve three-ring binders, one for each doctor, and see who might have an open spot."

Tap-Tap-Tap

Reflecting on our quaint yet mildly horrifying past, I'm relieved that our current EHR allows *tap-tap-tap*: instant access to any patient record.

Yesterday, for example, my patient called me about a new diagnosis of cancer. Today, I was able to read her oncologist's note, review the tissue pathology, refresh my knowledge about peritoneal carcinomatosis in an online textbook, secure chat with the oncologist about prognosis and treatment, set up a video visit with the patient and her family, and have a have a well-informed, thoughtful discussion about her next steps.

This would not have been possible in the era of paper charts.

The Lesson Learned

As our profession changes, it is natural to look back fondly at the "golden years of paper." They were not so golden. Disruptive change is coming, and your historical perspective will balance how you handle the coming years.

1994, Our First EHR. What Could Go Wrong?

Already in trouble, the paper chart system got steadily worse as our organization grew and spread to more buildings. We grumbled about the inefficient charts, but we were resigned to them. Our proposed solutions showed constrained thinking: "Let's buy more fax machines. Let's hire more medical records staff to reduce our two-week backlog. Let's hire more couriers and deliver paper *four* times a day instead."

These ideas were considered too expensive and unlikely to help. In the meantime, other hospitals in town were trying out electronic records.

The Veteran's Administration Hospital (Denver VA) installed CHCS II. In the mid-1990s, doctors there were already typing their progress notes and could easily look up lab results. Doctors' notes went from illegible handwriting to (cryptic) typed medical abbreviations: "Timolol 2 gtts OS BID. IOP 16OS/14OD." (Translation: "Two drops Timolol administered to left eye twice daily, Intra-Ocular Pressure 16 left eye, 14 right eye.")

To solve the paper delivery problem, Denver Health was scanning images of doctors' scribbled notes. The scribble was hard to read, and the low-resolution scan made it even harder. Neither system was popular with doctors.

In 1994, the University Hospital leaders heard about electronic records and saw an opportunity to improve things. They installed the 3M Clinical Workstation (CW). To the doctors, this was a revelation! Once lab results were connected in 1999, we no longer deleted results after 90 days. We could store and track them... *forever*.

However, CW had no scanning module, no easy way to record progress notes or medication lists, and no way to print prescriptions. Ultimately, it became a lab results repository. Everything else remained on paper.

Still, it was exciting. We had computers and a central place to store long-term data! We had a team: a handful of IT programmers, 50% of the time of one nurse and 20% of one physician (me).

This excitement unfortunately did not carry to the entire hospital.

After we installed CW in two clinics, how did it go? How did busy doctors like using it? Short answer—they didn't.

The primary care pilot site found CW to be only mildly useful. "It's slow, but it is nice to see recent test results, especially if paper charts are missing."

This lackluster encouragement was actually one of the better responses. The surgery clinic site was plainly uninterested. I wanted to install a computer and internet connection and show them what it could do, but surgery leaders declined. "This computer might be useful in primary care, where you write prescriptions and order lots of tests, but surgeons don't need this. We do surgery. We help people."

I reported these findings to the leaders: lukewarm acceptance at best. Fortunately, they did not cancel the program outright. However, instead of expanding the program to more clinics, as originally budgeted, they chose to allocate the bare minimum to "keep the lights on." Their instruction to our team: "Sounds like this is not that helpful. However, if you find other clinics that are interested, we have some discretionary funds to buy a few more computers. Maybe focus on primary care?"

With this, the fate of CW became clear: I didn't have strong leadership backing, so it would be a patchwork of the willing. The project limped along for years. This was a painful chapter in my career. At the time, I knew nothing about change management. We had spent our money on hardware and software and paid almost no attention to people and process.

What might we have done instead? Knowing what I know now, I would have focused on test results and the challenges that surround them. The physicians struggle to find timely and longitudinal results, and CW could have provided them. This "minimum viable product" could have been a big win. I could have communicated the primary benefit. Instead, I tried to teach everything, including tools that did not work well. As a result, CW was only installed in a handful of clinics and hospital floors.

The Lesson Learned

Do not deploy an EHR because it has cool tools. Deploy it when it satisfies a clinical need. For example, 3M Clinical Workstation was great at "test results." A better INVESTIGATION would have led to a narrower focus and stronger message.

The Sonar Sensor Conspiracy

In 2005, our doctors faced a hodge-podge of computer systems. We would use Allscripts in clinic, McKesson in the hospital, 3M Clinical Workstation for other clinical notes, Outlook to send messages, and PACS (Picture Archiving and Communication System) to look at X-rays. You got started each day by typing in your five passwords, arranging your windows, reviewing all the data on your first patient, and then going to see that patient.

This worked great, assuming all your patients were on a single floor of the hospital. Set up your desktop, go see a patient, come back to your desk to review data on your next patient, go see that patient, and so on. But what if your patients were scattered across many floors of the hospital?

Before leaving one floor, you had to close all your windows. Then walk to the new floor. Find an available PC. Log in to five different programs. Review the data. Go see your patient. Close everything and change floors again. When you have 12 patients on eight floors of the hospital, this gets old fast.

The frustration was: "We would like to take our tools with us as we move between floors. Please, don't ask us to carry laptops. The screens are too small, they're infection risks in the patient rooms, and they get lost."

Our IT department's brilliant solution was the "traveling virtual desktop." The new process looked like this: Log in to your five programs. Then, when done at one location, push a button to "secure the desktop." Walk to another floor and find any available PC. Log in to your traveling desktop, and *voila*, all your windows come back to life, arranged just so. Even better, we had RFID (radio frequency identification) employee badges that would remember your password for 12 hours. We installed badge-readers at every PC. One badge-tap, and you were in! Magic!

In the early days, this was very popular. Clinicians loved the mobility. The five-login frustration became a single "tap in"!

But soon after, we ran into a problem. Our clinicians loved "tapping in" to the traveling desktop. However, many did not remember to "tap out" when walking away. This was not good for a couple reasons: it left login sessions open, with patient information visible on the computer for anyone to see. Worse yet, not "tapping out" meant you couldn't "tap in" at the next computer. You were locked out. You would have to walk back to the previous location, which could be several floors away, and log yourself out. Or worse, you'd call someone on the other floor and beg. "Hey. Find that PC I was on and log me out, please?"

Traveling desktop "tap in" was a huge efficiency boost, but now we had the next issue: how to fix the "tap out" problem?

IT to the rescue again. The fix was a tiny sonar sensor the size of a pencil stub affixed to a gooseneck metal stalk that plugged into a USB port. When plugged in and pointed in the appropriate direction, it could detect the presence of a person. The sonar could also detect when a user walked away and automatically secure the desktop. No "tap out" needed. Problem solved! And sure enough, the sonars worked. As users walked away, traveling desktops logged out automatically.

But then, some sensors stopped working. We sent technicians, who found sensors with masking tape over their lenses. Others had been unplugged. The techs would plug the sensor back in or remove the tape, and it would work again. Next day, the tape was back. Soon, sensors began disappearing. Some were found in the trash. What was going on?

Time for INVESTIGATION.

This time, the technicians went to the locations of the unplugged sensors and asked innocent questions of nearby clinicians. "Do you know about these sensors? What do you think they are for, and why are they being unplugged?"

"Oh," they heard from docs and nurses, "the Hospital is checking on us. Maybe we aren't working fast enough, but I don't want someone looking over my shoulder. They don't get to spy on me."

Suspicions can run deep, apparently. Belatedly, we realized that the sonar sensors did look like tiny cameras. We could understand why the misconception might arise. In response, we removed all the sensors.

A few weeks later, one of the technicians came up with another idea. If users were suspicious of the "camera" for spying, what if we changed its appearance? Sonar does not need high resolution. If we put a wire mesh over the lens, it might not look like a camera. After some fiddling, we installed the mesh and redeployed the sensors.

Success! These revised sensors were left alone. They didn't look like cameras, and there was no implication of spying. The walkaway automation worked. Problem (finally) resolved.

To solve the multi-login, multi-floor problem, we had to tackle virtual desktop tech, badge tapping, and sonar. That's just the tech. Then we had to recognize and address users' walkaway behavior and sensor suspicions to be successful.

> **The Lesson Learned**
>
> It is not the technology that determines success. One must INVESTIGATE and understand the user's viewpoint. Be willing to fail, listen hard, and try again. Sometimes a series of minor tweaks make all the difference.

Surgery Orders and Complexity: The Cynefin Framework

In 2023, the surgeon informaticist and I were asked to implement the electronic Case Request Order (CRO) and Informed Consent (IC) for 1,000 surgeons across our health system. By doing so, we would eliminate a well-established paper process. The CRO/IC process, when done well, reduces errors in surgery by ensuring the patient, surgeon, and operating theater staff are on the same page. Moving to a digital version could reduce the existing paper pitfalls creating missing data problems.

How does this work? A CRO describes the intended surgery. The surgeon specifies "knee replacement" with a specific CPT (Current Procedure Terminology) code. For example, they might specify the right knee, using the XYZ approach, and the appropriate equipment. The clinic scheduler selects a day and time, and a staff member contacts the insurance company for "preauthorization" to pay for the anticipated surgery. The operating room scheduler confirms that this time does not conflict with other surgeons and that the anesthesia team and the right equipment trays are available. CRO is the centerpiece of this delicate dance.

Informed Consent is similar. The surgeon describes the risks and benefits of the proposed surgery to the patient. If the patient agrees, they sign the form. That paper is stored and transported in a folder to the OR on the day of the surgery for the safety cross-check: correct surgery, correct site, correct equipment, patient consent, all team members present.

CRO and IC involve many people. The process is well accepted but easily derailed. It has many steps including a paper checklist, several crucial forms, and the timely delivery of plastic folder to the right room for each of 100 surgeries a day.

The naïve informaticist or health system leader may believe that this process is ripe for improvement (correct) and that surgeons and their teams cannot wait to try something new and possibly better (incorrect).

Despite the known problems with paper, surgeons and their teams are resistant to change. "Nothing is wrong with how things work now. We are not interested in any new electronic process."

Per Shakespeare, "[We] rather bear those ills we have / Than fly to others that we know not of? / Thus conscience doth make cowards of us all." Classic, and still true.

How did we get here? Our organization started using electronic medical records in 1994. Those efforts focused mostly on general medical or specialty medical practices. Tools for surgical practices were less developed. As a result, surgeons rarely participated in our EHR work, and the CRO and IC remained on paper.

Gradually, CRO and IC tools in the EHR improved. By 2023, hospital leaders pushed to reduce medical errors in surgery: missing informed consent forms, missing pre-authorizations, mismatch of equipment trays to expected surgery, confusion about the planned surgery. The missing data led to surgeries being canceled at the last minute or, worse still, wrong-site surgery or other medical errors.

In 2023, our informatics team took action. We brought on a surgeon with informatics training to be a technical expert and we convinced surgeon leaders to be champions of the CRO/IC projects to improve surgical safety.

With such a guiding coalition of leaders, we were ready to take the first step by INVESTIGATING. We needed to understand the many forces at work in the surgery clinics and operating rooms that would float or sink this project.

The Cynefin framework, developed by Dave Snowden, is a useful way to think about complexity in organizations. Let's use it to dissect the CRO and IC projects. Cynefin describes a spectrum of projects from straightforward to difficult by defining Clear, Complicated, Complex, and Chaotic environments[1].

A **Clear** environment has straightforward and established rules. In isolation, a surgical case request order is "clear." For one surgeon and one operation (knee replacement) there are a small number of individuals involved and a limited set of data needed. It is straightforward to teach a handful of individuals how to use an electronic form and deliver the information at the right time so that the knee surgery can proceed safely.

A **Complicated** environment has multiple tasks, each with straightforward rules. The individual tasks are straightforward, but there are many interlocking parts that depend on each other. In the case of CRO/IC, adding 1,000 more surgeons and many more surgery specialties makes the environment **Complicated**. A careful leader who can lay out all the parts can achieve success routinely.

A **Complex** environment has multiple goals, but the rules are not straightforward. Some rules may be hidden and must be discovered. In leading a project, one must pursue goals while discovering the rules. Such emerging rules threaten the success of the project.

For example, a colleague proposed combining the informed consent for the surgery and anesthesia. Usually, these are separate consent conversations on different days. Is it even legally permissible to combine them? Is this best practice? Are there new risks? We must discover what the rules are. Five groups of people (surgery, anesthesia, legal, compliance, operations) will have to agree to a new process. Discovering and agreeing on the rules could take a long time. Any one group could stop the project entirely. Adding this proposal to the scope of the project could delay the timeline or stall the project completely.

In a **Chaotic** environment, no right answer is obvious. In such environments, even cause and effect are shifting. Fortunately, we weren't facing a chaotic environment. Had we been the first healthcare organization in the country to eliminate paper and adopt electronic CRO/IC, where the technologies had yet to be built, this project would have been operating in a chaotic environment.

With CRO/IC, we started with a complicated project: lots of moving parts, but the rules for success were known. If we added the combined consent sub-project, the rules were NOT known, and important stakeholders held unknown opinions.

In Complex environments, leaders must watch for emerging threats to success. There is no "best practice" for Complex situations. One must "probe, sense, and respond." For us, this means having early conversations (probe), determining whether there are many objections to the proposed combined consent (sense), then choosing whether the risk to the project was too great to add this additional objective (respond).

In this case, we agreed that the risks were too great, and we removed the combined consent sub-project. Furthermore, we also agreed to separate the CRO and IC projects and tackle them sequentially. These two adjustments shifted this project from Complex back to a more manageable Complicated environment.

Projects in informatics run the gamut; they can be clear, complicated, complex, or chaotic. Sometimes the informaticist has no choice and must simply push forward the best they can. With INVESTIGATION, it is possible to shift a project away from chaos and complexity and choose a path with a higher chance of success.

The Lesson Learned

As da Vinci supposedly said, "Simplicity is the ultimate sophistication." Healthcare organizations are complex and often chaotic environments. The Cynefin framework is a useful way to think about projects. When you can, nudge your projects away from chaotic and toward clear environments to reduce your risks and improve your chance of success.

INVESTIGATION—All the Lessons Learned

Intro—The Frog at the Bottom of the Well

Balance the parochial with the wider view. The passionate frontline clinician has a good local idea. Excessive variation is error-prone and inefficient. Both are important. Raise the frog from the bottom of the well so that their viewpoint changes and you work together. Understanding how clinicians think, and what came before, and how the larger system fits together, is crucial to the difficult task of introducing a new technology.

Stat-LAN and the Chief Complainer

Listen first. It is surprising how often we don't. Assume your colleague or patient is thoughtful. "Kids have ears and no mouth." Reflective listening is your superpower. To paraphrase a colleague: "The more I listen, the more my informatics projects succeed."

How It All Started

As our profession changes, it is natural to look back at the "golden years of paper." They were not so golden. Disruptive change is coming, and your historical perspective will balance how you handle the coming years.

1994, Our First EHR. What Could Go Wrong?

Do not deploy an EHR because it has cool tools. Deploy it when it satisfies a clinical need. For example, 3M was great at "test results." A better INVESTIGATION would have led to a narrower focus and stronger message.

The Sonar Sensor Conspiracy

It is not the technology that determines success. One must INVESTIGATE and understand the user's viewpoint. Be willing to fail, listen hard, and try again. Sometimes a series of minor tweaks makes all the difference.

Surgery Orders and Complexity: The Cynefin Framework

Healthcare organizations are complex and often chaotic environments. The Cynefin framework is a useful way to think about projects. When you can, nudge your projects away from chaotic and toward clear environments to reduce your risks and improve your chance of success.

INTRIGUE

Once you're truly improved your listening skills and have some sense of what your clinical and IT colleagues have to say and the work that they do, how do you get them interested in what you have to say? How well do you communicate? Which is more convincing: data or story? How do you intrigue your peers and get them on board?

Intro—If You Build It, They Will Not Come

In the early 2000s, clinics and hospitals began installing electronic health records to track important patient data. Unfortunately, most of these EHRs were stored within a single organization. Obtaining records from another organization was difficult.

A few years ago, one of my patients arrived in clinic to tell me "I was in the hospital for a few days across town. I had pneumonia and was pretty sick. While I was there, they found something in my lung that I should tell you. I don't know what it was. Also, they gave me a new medication that they said not to run out of. I ran out."

A quick search of my EHR found no mention of her hospital visit, since the other hospital was on a different computer system. Then I called the operator to find the number for medical records at that hospital. They were closed after 5:00 pm. I called the next day to request records. I faxed them a patient-signed "release of information" request. Several days later, I received a half-inch of faxed papers to read. Yes, she had a new lung nodule suspicious for cancer. Yes, she was put on a heart medicine and a depression medicine. This was several hours of work spread across my clerk, my nurse and myself, and was typical of how hard it is to move patient data between hospitals.

In other words, it was an ordeal. And as a result, doctors and nurses often treated patients without all the data during the clinic visit. OR they must spend time to assemble the information and finally call the patient with updated advice days later. Worse yet, doctors would sometimes just "repeat the test" to get the necessary information and make a medical decision quickly.

About this time, Colorado developed a Health Information Exchange. This was in 2009, during the national push for electronic health records driven by the HITECH act (Health Information Technology for Economic and Clinical Health).

Leaders of CORHIO (Colorado Regional Health Information Organization) built a database that connected patient health records across the state to reduce medical error and duplicate testing. The CORHIO challenge was that hospitals had grave privacy and security concerns of giving up control of their own data. They would have to trust that the copy of patient data that CORHIO obtained would be defended adequately from hackers. Therefore, the database had to be federated—a difficult technical design.

A federated database is a system that contains multiple databases, but they work together like a single database. Each hospital could keep their data secure, and CORHIO would set up a fast-access central hub that would quickly and securely link a requestor to the original data across many hospitals. This cooperation technology is the federated database. To the user, it feels like a single source of truth.

The technical achievement was a tour-de-force. The team launched a website called "CORHIO 360" for doctors and nurses. "Instead of taking hours to days to retrieve important patient records by calling and faxing, you now have instant access to medical records across the state."

What happened? Naturally, the trumpets blew, the walls came down, and the care of patients in Colorado improved dramatically overnight.

Right?

Sadly, no. Of the tens of thousands of physicians and nurses in Colorado, fewer than 100 signed up for a CORHIO 360 login.

What happened?

As I've learned repeatedly in my career, "If you build it, they will come" is a nice movie slogan, but it's a terrible informatics strategy. CORHIO had focused on the tough tasks of assembling data and sustainable funding. They had not spent enough time thinking about workflow: how and when our busy frontline clinicians would use the tool.

A busy emergency department doc may be treating four or more seriously ill patients at a time. They barely have time to login and read patient data in their primary EHR, much less, log out, open a web browser, login to CORHIO 360, type in the patient's name and date of birth "in case this patient has relevant data at other hospitals." This could take three or more minutes per patient.

In busy emergency departments, every second counts. Our cardiologists say, "time is myocardium." Meaning that every minute delayed is heart muscle dying during a heart attack. "It's just another webpage and login" may be reasonable for an office worker sitting at a computer, but it's it is a completely unreasonable request for a doctor racing to keep a patient alive.

Hence our twist on the movie slogan "If you build it, they will not come." It doesn't matter whether your tech is amazing. If it is not easy enough or quick enough, docs won't use it, and then nothing else matters. Similarly, we also say "Workflow eats technology for lunch."

Some motivated teams got creative and empowered clerical staff to pre-fetch CORHIO results on high-risk patients while they were still in-transit via ambulance to save time for doctors and nurses. INTRIGUED teams will find a way to incorporate useful tools.

In later years, the IT teams found a way to automatically merge CORHIO 360 data right into the primary EHR so no separate login was needed. Doctors found the data in their usual workflow. Our cross-hospital use of data improved greatly after that.

The Lesson Learned

Your technical achievements may be great, but you will stumble unless you INTRIGUE your clinicians and interest them in redesigning the workflow. "If you build it, they will not come." Remind yourself that tech alone is not enough.

What Is a Sacred Bundle, and Why Do You Need One?

The Pawnee and Lakota Native American tribes once used bundles of tokens to represent their culture as they traveled from place to place, following the buffalo migration. At every new settlement, the elder would gather the tribe and present the items one by one. Each token was accompanied by a story: "This stone was from the river where our tribe first..." Then, "This feather was from the headdress of our bravest warrior. He died protecting..." And so on. Each time these stories were retold at each new location, that settlement became home. The values and lessons of these stories ran through the tribe, and all members came to know them by heart. Writer and organizational guru Peg Neuhauser[1] referred to that cloth sack of cultural tokens as a "sacred bundle."

Why is this important? Why is it relevant to informatics? When our teams were small, everyone fought the same dragons: physicians resistant to change, a paternalist tradition of not sharing data with patients. Without this common experience, it would be easy for our technology choices to drift, sometimes favoring the old ways.

Our newest users may insist "I would like to receive test results by fax instead of by EHR In Basket." Or "I don't want any of my patient's test results automatically released to the portal. They don't need to see them."

As we added informaticists to our team, the newbies did not yet bear the same battle scars, the common experience of pushing boulders uphill against walled cultures of inertia and information secrecy. Without knowing the "why", they might inadvertently allow users to return to older ways.

Perhaps a sacred bundle of key stories would help. Telling stories is a vivid way to teach our teams common purpose and vision.

We tell our clinical and IT teams the sacred bundle story of the OpenNotes and Open Results journey (see **The Evolution of OpenNotes** on page 184). This journey established the cultural norm of sharing ALL records with our patients. Retelling this story reminds our teams of why we do this important work. The story includes the naysayer showstopper: "CT, I don't know where you went to medical school, but where I trained, the medical record is a *dangerous place*. Patients have no business reading their own records."

We tell the story of **APSO notes** and the effectiveness of the "meeting before the meeting." (see **APSO Notes, Maxwell, and the Meeting Before the Meeting** on page 98). It teaches us how much harder the psycho-social politics of physician adoption are than the underlying technology innovation.

We tell the story of Sprint teams (see **Sprint Informatics** on page 105) and how it took three full years to go from innovative idea to first trial run, and how "stealing a team" was what it took to get started. Persistence, sneakiness and luck helped us succeed.

These stories illustrate the importance of information transparency, psycho-social political skill, and persistence and luck. These values are deeply rooted in our informatics culture.

Collecting these stories, however, is only the first step. In order to be effective, the stories must be shared and known among your team. Find colleagues who can bring the stories to life and share them regularly until any team member can tell them.

In a larger context, this book is my sacred bundle. I want to pass on to you—current and future informaticists—the way we gathered a team, established a culture, and built trust to lead difficult changes in a healthcare organization.

I have been toying with the idea of making a physical cloth sack with sacred bundle objects representing the key stories of our organization. Perhaps I would pack a rubber boot: an artifact that represents my Chief Complainer days of stomping angrily and demanding answers without being a good listener. Or perhaps a sea-going oil rig, a representation of a classic burning platform from which users must jump when change is called for.

From these stories we reflect on what is important. Then we relate them to our current projects, and how we are each facing setbacks, and drawing encouragement from our most important values, elevated by our sacred bundle stories.

> **The Lesson Learned**
>
> Stories make up the DNA of your organization. Your team should know them. Collect and tell the stories that intrigue your team and build your culture. What stories do you tell? What objects go in your sacred bundle?

A Surgeon's Friend

How do you communicate with colleagues who don't read emails or attend meetings? One of my colleagues told me, "The more I listen, the more my patients learn." It took me some reflection to understand this non-obvious statement. A physician who is a good listener understands their patient deeply, including what is important to them, and what they are willing to change. Then, the physician can shape their advice to what the patient can hear, learn and do. Listening is a superpower.

In the mid-2000s, as Medical Director of Informatics, I began sending lengthy EHR update emails to all medical staff—about 900 physicians and APPs. I patted myself on the back, satisfied that I was keeping my colleagues up to date on EHR topics and explaining why we were making changes.

It was Reenie Zaccardi, VP of ambulatory care and my good friend, who dropped by to tell me: "Surgeons don't read past the first sentence of an email, if they read at all. Your internist style of long explanations is not getting through."

This was a shocking idea. I was sure everyone had been hanging on my every emailed word. Just kidding. I realized that she was right: in the age of information overload, no one reads email carefully, and surgeons in particular have no patience for messages that aren't quick and clear.

I am, after all, the same person who wrote a 17-page white paper that addressed every possible objection my colleagues might have to Open-Notes and sharing medical information online with patients. The result? No one read it.

I recalled a book I read in high school, *The Elements of Style* by Strunk and White:

> *Vigorous writing is concise... This requires not that the writer make all his sentences short, or that he avoid all detail and treat his subjects only in outline, but that he make every word tell.*

Starting that day, I set myself a goal of writing pithy emails. I would keep them as short as possible, as inspired by Strunk, but also in the style of Poor Richard's Almanac, a collection of brief tips for farmers. I decided to write under the pseudonym "A Surgeon's Friend" to make fun of myself, and I made myself write only one EHR tip per email, in as few words as possible. I never cut things down to a single sentence, but my emails were dramatically shorter. For example:

> *SUBJECT: A Surgeon's Friend: Use Magnifying Glass to Simplify Searching in TouchWorks*
>
> *Ever wonder if there is a faster way to find the CT or MRI in a chart? There IS! In the patient's chart, click on the Magnifying Glass icon. Type "MR" to find all MRIs. Type "CT[space]" to find all the CT scans. "oper" finds Operative Reports. "clin" finds all clinic notes. Wanna learn more? On the TouchWorks main screen, click on "Tips and Tricks."*

Did I receive even one appreciative email reply from my hundreds of surgeon colleagues? No. Did I get Reenie's attention? Yes. She also noted that, listening in on surgical conversations, that "A Surgeon's Friend" was being noticed and discussed.

> **The Lesson Learned**
>
> People are constantly telling you how well you communicate. Listen and look hard. If you're not getting through, it's on you. The responsibility of communicating lies with the sender. INTRIGUE them. A simple message carries more impact.

What TikTok Teaches Us About Communication

Oldsters like us, take heed: TikTok and 15-to-60-second videos are *in.*

TL;DR: Nobody has time for anything. But as demonstrated by the popularity of short-form videos on Instagram, YouTube, and TikTok, if you make it clever and interesting, users might take 15 to 60 seconds to watch your content.

TikTok, started in 2016 as a lip-syncing app. Although recent iterations of the app allow for videos up to an hour in length, the initial short-form limitations force the creator to get their idea across in 15 seconds or less. This 15-second time restriction began as a concession to the music industry's maximum replay length of a copyrighted song.

I already had a YouTube account with several videos when I downloaded TikTok. I experimentally posted 60 seconds (then the maximum length) of one of my ukulele song parodies and could barely believe the response. Unlike my languishing YouTube videos, whose view-counts slowly climb into the double digits, my TikTok views leapt up to 1,000 in a single day.

"Wow," I thought to myself, "I am *amazing* at TikTok." (For those interested, my TikTok is linked in **Appendix 2**.)

But no, this is just the TikTok algorithm putting your video in front of a broad range of viewers, no channel subscription needed. It's common for videos to garner 1,000 views in *an hour*. The hashtags and close-knit social media communities we had grown over time were deemphasized in favor of instant, superficial attention and the allure of the endless scroll.

And yet, it is still possible to convey complex ideas, tell a story, and change minds in just 15 to 60 seconds. See: chefs sharing recipes on "CookTok" and readers describing new books on "BookTok."

In contrast, our team at UCHealth just retired a 17-minute video I made a decade ago. It gave a full walkthrough of how to use the EHR for a clinic visit. Looking back at it today, I can't help but cringe a little. Whew, how out of touch was *that* guy?

Our more recent videos are much shorter—just a minute or two—and focused on a *single* technique or tool. Those viewers who quickly "get it" can be done watching in 15 seconds, but the more interested viewers can pause to catch subtle details. And the video automatically loops, so the viewer can enjoy your jokes again. Hmm, is this a paradigm shift? Should we embed TikTok-length educational videos into our EHR?

Put Road Signs on the Highway

As we say internally, shouldn't we put the "road signs" (our tips and tricks) on the "highway" (the EHR itself), and not in the "garage" (our email newsletters and online reference library)? Aren't our users more likely to click on tips when they're working? Not when they spontaneously think, "Oh, I have some time, let me see what I can learn." Which is never.

Austin Chang Is My Hero

There are doctors who have mastered the short-form video. For example, Harvard gastroenterologist Austin Chang.[1] He teaches good medical advice in just 15 seconds of hilarious music and terrible dancing-in-scrubs.

Dr. Glaucomflecken (a character created by Dr. Will Flanary)[2] uses short videos to poke fun at medicine and the business of medicine with his multiple personalities, all in less than a minute. Brilliant.

Seeing these talented creator/educators, I am both pleased and frightened. I bemoan the deterioration of the general public's attention span, but I am excited by the new idea of shoehorning valuable education into humorous tiny bites in this format. (Inspired by this content, my team created a TikTok-length video on the responsible use of Epic Secure Chat we called "Haiku Hero." See **Appendix 2**.)

> **The Lesson Learned**
>
> Attention is short. Videos should teach, entertain, and be no more than a minute or two. As agents of change, and educators, how will we use this?

Paternalism and the Patient Portal

Paternalism is deep in the DNA of medicine. Even Hippocrates' original oath (when translated) states: "I will impart... whatever I know... to my master's children as to my own and likewise to all my pupils... but to none else."

Paternalism tells us that the physician knows more than the patient. It says that sometimes, it is best that medical diagnoses and decisions go unexplained.

In contrast, patient advocacy groups in recent decades have been outspoken that doctors should share more information with their patients. We believed this was the right thing to do, but we struggled to convince our medical colleagues to move from paternalism toward transparency. (For our full journey, see **The Evolution of OpenNotes** on page 184.)

One researcher, Ed Krupat, published a paper on the Patient-Practitioner Orientation Scale (PPOS)[1] that shed some light on our struggle. His paper described a brief questionnaire that could score a patient's attitude as being "Patient-centered" or "Doctor-centered." Questions on the instrument ask respondents to rate statements on a scale from "totally disagree" to "totally agree." For example, "It is often best for patients if they do not have a full explanation of their medical condition," or "patients should be treated as if they were partners with their doctor, equal in power and status."

Using the responses, one could categorize both patients and doctors into the two groups.

In Krupat's 2000 study, patient and doctor pairs completed the questionnaire, and then the patients rated their satisfaction with the visit. Some findings seemed obvious: visits between patient-centered doctors and patient-centered patients, with aligned attitudes, resulted in high patient satisfaction. Similarly, visits between doctor-centered doctors and doctor-centered patients also resulted in high patient satisfaction.

Some of the discordance was also easy to predict: Visits between patient-centered patients and doctor-centered doctors resulted in low patient satisfaction.

But what about doctor-centered patients with patient-centered doctors? How satisfied were the patients?

Surprisingly, they were "very satisfied."

	Patient-Centered Patient	**Doctor-Centered Patient**
Doctor-Centered Doctor	Not Satisfied	Very Satisfied
Patient-Centered Doctor	Very Satisfied	Very Satisfied

Krupat's results lead to several ideas. The first insight is that, surprisingly, not all doctors are patient-centered in their attitudes. Some are clearly disciples of Hippocrates' original statements and take a doctor-centric stance.

The second insight is that patients are universally satisfied with the patient-centered doctor, regardless of the patient's own attitudes, whereas the doctor-centered doctor only gets high marks when the patient happens to be doctor-centered as well. In my experience, patients' attitudes have shifted to be much more patient-centered in recent decades.

In treating disease, the patient-centered doctor ensures that a patient feels heard, that data is shared, and that decisions are made jointly rather than unilaterally. Regardless of the patient's own attitude (patient- or doctor-centered), they appreciate a doctor who listens closely and adjusts their approach to each patient.

These patterns extend to information transparency. Doctors with a doctor-centric attitude may also be against the idea of sharing progress notes or test results with patients. This is the reason we struggled with our patient information transparency projects. Older doctors, those in leadership positions with decision-making power, were more likely to have doctor-centric attitudes and less likely to support information transparency. Finally, the patient-centered approach can be applied to the relationship between informaticists and our EHR-using colleagues. When possible, allow EHR customization by personal preference. Being patient-centered, or in our case, "user-centered", leads to greater satisfaction. Where you can, dismantle informatics paternalism: "We know better than you. Just trust us." INTRIGUE your colleagues and you will be more successful.

The Lesson Learned

Unwind paternalism toward patients and toward EHR users. A patient-centric attitude makes a better doctor. A user-centric attitude improves informatics.

Haiku Poetry

Why use APSO notes?
Putting the conclusions first
makes reading easy.

I have been sending a monthly newsletter to my colleagues for years. It can be tough to break through the noise of constant interruptions, messages, and emails.

At one point, the open rate for my newsletter email was about 12%. Fewer than 1 in 8 colleagues read our news. The reason is not hard to intuit: We all have information overload. Just like our patients tell us they were "too sick to come to the doctor," my colleagues say, "I'm too busy to stop and learn how to be more efficient." How can we educate those with no time to learn?

What's alert fatigue?
Too many alerts lead to
docs who "ignore all."

With more media to consume than any one person could possibly manage, ours is an "attention economy." Everything competes for your time and attention. Everyone has ever more emails, more alerts, more interruptions in the day. As my mother would say, "Aiyaaa." What is one to do?

Epic can do THAT?
Huh. Now we must figure out:
Is that a good thing?

What will break through the email habit of: "Scroll, delete. Scroll, delete"? Thinking about this problem, I noticed how many of my terminally busy colleagues scroll social media on their phones. I notice in myself how effective clickbait is in social media. If cleverly written clickbait gets attention, why not write clickbait-like clever haiku for the public good?

Careful you must be
to never let another
doc use your password

I briefly tried writing longer form poetry. Not my style. Even hokey limericks were too difficult. However, I found the 17-syllable haiku approachable and fun.

Can I still dictate?
Of course, but more clicking means
safer patient care

I started adding a haiku to the beginning of every Epic Provider Newsletter. It seemed weird but poetic to write a haiku about Epic.

Footsteps approaching:
Just more Epic analysts.
You are not alone.

One provocative haiku I never published in our newsletter, but might be true:

Even surgeons can
place their own orders online.
Fly the pigs will, then.

Newsletters don't often get replies. But when I started including haiku, I got replies like:

"Nice haiku."
"Your haiku syllable count is wrong."
"Your poem isn't really a haiku."

How funny. In trying something new to generate interest and response, most of the responses were critical. Aiyaaa. However, I take heart that to criticize, my colleagues had to actually open and read at least part of the newsletter. Success!

I have persisted writing haiku about Epic. In the intervening years, I regularly got supportive comments about my haiku. More to the point, our newsletter readership is up to 50%. Perhaps an introductory haiku does not hurt.

> **The Lesson Learned**
>
> A few syllables
> Attention, whimsy, meaning
> INTRIGUE with poems.

The Superpower That Is the One-Pager

No one ever made a decision because of a number. They need a story.

—Daniel Kahneman[1]

A good idea, ignored, helps none.

It is tough to catch your boss's attention. They are busy executives in a large health system with many worries: delivering high-quality care, addressing patient complaints, finances, and cultivating a public reputation. When you have an idea for worthwhile change, it often requires funding, and funding requires executive approval. That means making a pitch. That's the challenge.

I have seen many funding pitches in the boardroom. Many were accompanied by slides in the standard format of "Six bullets per slide, six words per bullet." Many had confusing tables and graphs. As a result, I saw many proposals fail.

This phenomenon is not unique to healthcare. It is likely that an engineer's unclear slides and sentence fragments contributed to the Columbia shuttle disaster that killed seven astronauts[2].

If a standard slide deck is not effective, what is?

As a meticulous internal medicine doc, I am great at sweating the details and explaining them comprehensively. That is one of my strengths. Pick a topic, and I will research it deeply and then write you pages of explanation or give an hour-long talk about it.

Unfortunately, this is *not* the skill you need as a leader. Knowing too much makes it hard to speak clearly. Being an expert is not the same as being an effective advocate. In a boardroom presentation, you only have a minute to make your point. I believe that leaders often hear good ideas, but because the presenters are not clear and brief, the opportunity is lost.

When building a Sprint EHR optimization team (see **Sprint Informatics** on page 105), I had the task of convincing leaders to fund this program for clinic doctors and nurses. I found it difficult to explain the purpose of this paradigm-shifting idea: Instead of teaching EHR skills through cheap email newsletters, we should build a team of doctors, experts, and trainers to do a deep immersion, one clinic at a time. I believed deep immersion was the only way to change EHR and teamwork in clinic. (Turns out, I was right!)

In the days following my failed, overly detailed boardroom presentation, I tried a new tactic. In the swirl of discussion over "What is this 'Sprint' everyone is talking about?" I set myself the task of writing a one-page summary describing the project, the pre- and post-Sprint data we collected, and the quotes and stories of people benefiting from Sprint. It was an easily shared, quick read designed to open minds. Some of the best quotes from my one-pager: "You made me a better mother. I get home in time for dinner with my family," and "I no longer contemplate early retirement to escape the EMR." Another memorable response was, "It feels like this organization cares about us."

The Sprint team did eventually get funded, and I believe the one-pager played a significant role in communicating its importance. Because of this initial success, my colleagues and I have used this one-page format many times.

Writing a single-page distillation of your idea is challenging. You have to be clear. Be brief. Combine data and story. Make hard choices to eliminate unnecessary details. Make it visually interesting. Know your audience. Know what major point you want to convey (you only get to make one point). And no matter what, stick to one page. This one-pager is far superior to printing out a slide deck, six slides to a page.

Here are 10 principles of an effective one-page summary.

- Keep to one side of **one page**. This forces clear thinking and concise writing.
- **Use a catchy title**. Capture attention.
- **Start with "why"**. Lay out the entire proposal in under four sentences.
- **Show data** for the left brain and **stories and quotes** for the right brain.

- **Know your audience** and their goals and concerns. It's not about you.
- Use **SBAR format**: "situation, background, assessment, recommendation."
- Include **visual interest**, pictures, illustrations, graphs. Avoid dense text.
- For a **project summary**, show a strong conclusion.
- For a **project proposal**, be concrete with your ask.
- A one-pager must be **understandable on its own**.

The Lesson Learned

One page is the perfect size to INTRIGUE your boss. Is it a report? Include data and stories. Make it brief and clear. Is it a request? Define the problem, have a clear ask.

(For example of specific one-pagers, see **Appendix 3**.)

What is Your Boss's Learning Style?

The burden of communication is on the sender. I learned this the hard way, so let me say it again: The burden of communication is on the sender. Really. You must INTRIGUE your recipient. And recipients are not all alike.

My past bosses have had very different learning styles. Some liked to talk and brainstorm new ideas during meetings. Some ran short, decision-focused meetings and needed time to digest new ideas afterward.

I am often in a position to propose a new idea or project, and so I am also a frequent recipient of the ensuing "No." After once again bandaging my wounded pride, I sit and rethink my approach. When an idea gets rejected, it isn't always the fault of the idea.

In my earlier career, I had a boss who was flexible in meetings and would welcome open debate and new ideas. We could discuss and brainstorm freely, then craft a potential project. We enjoyed the creative interplay of ideas. I learned that I could expect my new ideas to get a fair hearing.

If something occurred to me before the meeting, I would sometimes propose a new idea via email. However, the method backfired. In many cases, it provided time for my recipients to find reasons for rejection. As a result, I learned to bring ideas to meetings and not send them ahead of time.

Then, with a different boss, my focus on during-meeting proposals stopped working. Bringing up new ideas in a fast-paced, highly structured meeting often resulted in a high-pressure, superficial discussion. It was frustrating, as I struggled to articulate my position quickly and then defend it. The answer was often a reflexive "No, I don't see why we should do that."

In some cases, I persisted after the meeting and followed up with supporting data and a more clearly written argument. A day later, I would receive a reply: "I read your writeup, and I changed my mind. I think we should do it." I realized that my boss disliked being surprised with new ideas during a meeting. He liked to have time to consider all sides of a topic.

In response, I reversed my tactics: Send a written proposal ahead of time.

My email might say something like, "I am sending this one-page summary for our meeting tomorrow. I have laid out risks/benefits/costs/alternatives." The next day, we would sit down for our meeting, and my boss might open with, "I read your proposal. Looks reasonable. Approved. Next topic." Amazing. On other occasions, it was "Don't need to discuss. We can't fund it this year. Good idea, though." Even then, I felt more satisfied that my proposal had gotten a fair shake.

Who knew that different leaders have different decision styles? Even though I knew that the burden of communication is on the sender, I did not *behave* like I knew this.

In meetings I hear people say, "You still don't get it." Or, "Like I said before..." When I hear this now, I cringe inwardly. In repeating ourselves, we are failing to adjust our communication styles when we recognize our recipient is not getting our message.

Your boss is the most important recipient of your communications. Without their understanding and agreement, your projects are going nowhere fast.

> **The Lesson Learned**
>
> Communicating with patients, colleagues, and direct reports is important. It is even more crucial to communicate effectively with your boss.

Genius and Ninja Stickers

Fun and effective *can* be cheap and easy.

In 2005, we were deploying the EHR in clinics at the university. We needed to get our users up to speed, but it was difficult to grab the attention of busy doctors and nurses who cared for a constant stream of patients. Who has time to pause and learn how to use a new computer system?

Most docs thought of the EHR as a nuisance. It was just a fancy way to store paper records—those things you had to face at the end of the day, where you would scribble some notes and then bind them in a folder so that you could remind yourself about the patient at a later date.

Our job was to shift that perception and improve our colleagues' EHR skills. We would help them find information faster, make better decisions, and prescribe meds and order tests more effectively.

About that time, Apple Computer was rolling out retail stores. Every Apple Store had a Genius Bar. Bring in your iPhone or iMac, and the "geniuses" would have a look and teach you or fix something and make things right. It was nice to get in-person, individual guidance for my problem from an expert. In short, Apple geniuses were cool.

This got me thinking—what if *we* were cool? What if we built a Genius Bar to help people with the EHR? We bought a stainless-steel rolling operating theater table, brought a couple of laptops, printed some tip sheets, and labeled it our "EHR Genius Bar." Too busy to pay our team a visit? No problem. If you won't come to us, we'll come to you. We rolled the table right out into the clinic hallway. Surely, someone would stop by to learn a trick or two.

It did not work out that way. We introduced ourselves to the clinic leaders, doctors and nurses, then parked ourselves in the hallway. All day long, busy patients and docs walked right by us, in and out of exam rooms and offices. Here we were, with our bright smiling faces, ready to help! And yet, almost no one stopped to talk.

This is when we realized we were battling more than just inconvenience. These busy clinics had tight schedules and many patients, and no one felt able to spare even a minute of "down time" to pause and learn something. How might we get their attention?

Our solution? The Genius Quiz.

Are you an *EPIC* Genius?!

General Epic

1. ***NEW!*** What are Haiku and Canto, and when will they be available for general use?

2. How do I use a PC in the Exam room most effectively?

3. How do I use the badge reader most efficiently?

4. When using a non-badge reader PC, what's the quickest way to log back in to Epic? (hint: passthrough)

Reviewing a Patient's Chart

5. ***NEW!*** What's the fastest way to find microbiology and pathology reports?

6. What's the best way to find notes that YOU or YOUR specialty colleagues wrote on a patient? (hint: filter)

7. When looking through a chart, how could I 'save' notes and results to quickly view later, like when I'm writing my note? (hint: sidebar)

Armed with a big new sign, we headed back out. "TAKE THE GENIUS QUIZ. Are YOU an EHR GENIUS?" Now we had their attention. Soon, people were stopping to read the sign.

"What is that?"

"It's an EHR quiz. If you get more than eight questions right, you're a genius. You get a 'genius' sticker."

"Let me see that."

Every clinician believes they're a genius. And even a busy clinician, it turns out, can spare a minute to prove it.

"Yes, I know that one. Yep. Yep. Wait, what is this?"

In taking the quiz, everyone found something they did not know. Gotcha! Once we had their attention, they would often sit for a few minutes and learn something despite themselves. We could talk them through a tip sheet or show a quick video, and they would walk away and immediately try something new in the EHR.

With our genius quiz, we regularly had visitors. If they got more than eight of ten questions right (with our help), they were an EHR GENIUS.

"What's my reward for being a Genius?"

"Here's a genius sticker!"

I started with an old box of self-adhesive address labels. I then found an image of Albert Einstein, a comic sketch of a brain, an image of Homer Simpson with a tiny brain, and labeled the stickers: "I'm an EHR GENIUS."

Phase One of the plan was complete. People were learning and having fun. The stickers were an instant hit. Got eight questions right? Here's a sticker. Even if we had to "clarify" a few things on your answers (by teaching you something you didn't know), we're pretty sure you got eight questions right. Here's a sticker! Every person who took our genius quiz walked away with a sticker—and a big smile.

Even better, as physicians and nurses walked around with flimsy address-label stickers on their scrubs and white coats, they got a lot of questions.

"That's hilarious, where did you get that?"

"Over there, go take the 'genius quiz.' They'll even tell you the answers."

Pretty soon, our stickers were everywhere in the clinic. They were stuck on laptops and desktops, on office signs, even on name badges. Word got out to other clinics. They started asking "When do *we* get a Genius Bar?"

Phase Two was complete. We had INTRIGUED our colleagues.

We rotated our Genius Bar to other clinics. We were engaging docs and nurses, delivering just-in-time education, and spreading joy. We basked in our success.

This success would not last.

When we brought the Genius Bar to our first surgery clinic, we were a bust. No one was interested. They just walked on by.

Why? What made surgeons different? Quietly, my surgeon informaticist colleague suggested: "Surgeons don't *care* about being geniuses. But ninjas? Ninjas are cool."

Bingo. Overnight, I created the surgeon-focused 'ninja quiz' and corresponding ninja stickers. We were back in business. We were just as successful with ninjas and surgery clinics as we were with geniuses and medical clinics.

The Lesson Learned

"Let me teach you something!" is less engaging than, "Are you a Genius?" And "I'm too busy" becomes "Thanks for bringing joy to our work." INTRIGUE your colleagues. Fun and effective can be cheap and easy.

The DOs and DON'Ts of an Effective Presentation

Don't suffocate the tender roots of your great idea. Effective presentations plant the seeds of your idea in the minds of others. Memorable talks increase the chance that your idea lands, and grows, and spreads.

As informaticists and physicians, we attend national meetings and get excited to hear great ideas and learn from others. If we are lucky, we are asked to give a talk. Unfortunately, too many talks are given by experts who have deep experience and insightful data but insufficient presentation skills. In other words, too many talks are *boring*. And therein lies the tragedy: Boring talks lose their audiences. Instead of spreading, their ideas die on the vine.

When experts give a talk, they often think, "I'll tell them why I'm great and why my projects are great." This becomes a dry checklist of accomplishments. After two or three awards are shown, the audience gets bored. "We get it. You're great." No struggle. No story. No one cares.

"I'm great" belongs on the trophy shelf in your parents' house, but not in a talk.

TED Talks show us another way. These talks are entertaining, they grab you from the first sentence, and they are visually interesting. Unfortunately, they are also light on data and nuance. TED Talks represent the other side of the presentation spectrum.

When I give a talk, I want both: academic, information-dense knowledge transfer *and* TED-like audience engagement. What could that look like?

One might think: Simple! Go download fancy presentation software, or ask a generative AI tool to create slides for you. BOOM! Done.

As Arthur C. Clarke said about teachers over 50 years ago: "Any teacher who can be replaced by a machine should be."[1] If you rely on these tools to make your talk, maybe you are not the human to give a talk. You should aspire to be better than generated templates. This is even more true with early generative AI content. I'm seeing white papers, slides, and emails that sound like classic business-speak. I worry, does human-to-human communication wither when we farm out our work to AI tools?

From my 30-plus years of giving and listening to talks, here are some ideas.

- Grab your audience's attention with the title and opening statement.
- Draw them in and lay out the struggle. What difficulty are we facing?
- Your slides will distract from your story and your wisdom. Don't let them.

Here are some paired DOs and DON'Ts. (Each row is explained in more detail below.)

DON'T	DO
1. Start with "Blah blah blah I'm an expert."	Start with a hook. Why should I care?
2. Speak in a monotone.	Vary your speed and volume. Talk to the last row!
3. Use 6 words x 6 lines so that your spoken sentences differ from what is written.	Write complete thoughts or quotes. Sometimes it's okay to just let the audience read!
4. Use video or tiny images.	Full-bleed images with story and emotion.
5. Give a boring list of accomplishments.	Show the struggle. Idea. Outcome. Story!
6. Distribute handouts of 6 slides per page.	Write a one-page summary of your talk.
7. Offer platitudes.	Offer specific, concrete lessons.

DON'T	DO
8. Be predictable.	Create surprises & interaction.
9. Leave your audience hanging.	Close the initial loop. Give a call to action.
10. Trail off or end suddenly.	End early. Give thanks. Control Q/A. Perform.

1. Stale Introduction. The standard opening required by conferences is: Pick a title to match the conference theme. Then describe your organization and its awards. Then describe your expertise, blah blah blah. I am already asleep. Instead, start with a hook. "Don't you hate it when...?" Or "Is this the end of...?" Or better yet, "We were facing a crisis!" Blast through mandatory slides, get to your content, your story. Make me care.

2. Monotone. Trying to sound official, too many presenters speak in a monotone. Insufficient practice is one cause: It's hard to sound engaged and active when you're reading the words off the slide. But then, too *much* rehearsing can lead to speaker boredom. Boredom is easily transmitted to the audience. Instead, infect them with your passion. Vary your speed and volume. Don't hide behind the podium if you can get a mobile mic and wander the stage. Use your entire body to convey your point. Speak to the person in the last row; you will get louder and gesture more. Use your body and voice to make your points BIG. When practicing, learn your slides well enough that you don't have to look at them, but not so well that you can remember every word verbatim.

3. The Ol' Six-by-Six. The classic advice is to use six bullet points per slide with six words per bullet. The rationale is that these 36 words fit on a slide and are easily read in large rooms. This is outdated.

Six-word sentences are not complete thoughts. They only make sense to the presenter. Write a complete sentence or an idea that is clear. Or, show one word or one big number and talk about what that means. If you wrote a quote or a complete thought, you can just let the audience read.

The opposite extreme is no better. In the post-pandemic world, we often watch presentations online. This has enabled tiny fonts and dense slides of text because presenters know the recipient is sitting inches from the screen. Yes, smaller fonts are readable, but this is objectively terrible. Even if I wanted a textbook, I would not want you to read it to me. I never want to hear "I know this is hard to read, but..."

Choose one idea per slide. One point.

Don't show a sentence and then *say something else*. The audience is made to choose whether to listen to you or read the slide. They will read the slide because they're worried they'll miss something important, which means they are not listening to you. The cognitive interference between your spoken and written words means they are absorbing nothing.

Take time to tidy up your slides. Make them clean. Don't mix fonts, don't allow "word widows" (automatically wrapped sentences that leave the final word dangling on its own line), don't overuse <u>underline</u> and *italics* and **bold**. In other words, don't emulate the Las Vegas Strip. Your ideas are the point, not your fancy set-dressing.

Don't put your organization's logo or the conference logo on every slide. Instead, use the space for your own purpose, and message. Restrict logos to first and last page if absolutely needed.

4. Stamp-Sized Images. Do not place a tiny picture on the screen, overwhelmed by your words. Also, avoid looping animations or video. It distracts from your point and is at risk of not working. Instead, add a full-bleed image (one that "bleeds" off the sides of the slide). If words are absolutely needed, add a few words in contrasting color over part of the image to make your point.

5. Litany of Accomplishments. No one wants to hear about a list of successes. Instead, tell us about your journey. Tell us the what, where, and how. What challenge were you facing? How did you approach it? Where did you struggle? Did you fail at first? How did you overcome that barrier? Pick one challenge. Put me in the room where you had an insight and began to have hope. Let me share your victory. One great story is memorable. A list of successes is not.

6. "Helpful" Handouts. Please do not print the standard six-slides-per-page handout. I have never seen a printed slide handout that was useful. Unless you've listened to the talk, printed slides make no sense. And, the resulting font is microscopic. Instead, if a handout is important, make the effort to write a one-page summary that can be read on its own or passed around to others.

7. Empty Platitudes. Too often, "lessons learned" are boring statements of the obvious. "Make sure your leaders agree before launching a big new project." "Don't underestimate clinician resistance to change." Duh. Take the time to write a concise but specific lesson. "Choose carefully: your clinical champion should be a respected leader, *not* the junior faculty computer geek."

8. Too Predictable. Insert surprises. For example, if you ask a question to the audience, print out a full page of that slide with the answer written on it, and hide it in the audience, taped under someone's seat. If no one has the answer when you ask, then instruct them to look under their seats, and ask whomever has that paper to stand up and read the answer. Or, you could insert polling questions during the talk and show results. You could ask audience members to pair off and share their answers to a question you pose. For example, "What was your most memorable setback in regard to...?" Or "Tell us one practical habit you have for productivity."

9. Left Them Hanging. Don't leave them hanging. Answer the provocative hook from your opening. Finish the hero's story. Call the audience to a specific action. "Now that you know this, go..." Summarize with one slide of concise yet concrete ideas. Pause for smartphone photos or screen captures if the audience finds it useful. Include a QR code linking to your one-page handout, if you wrote one.

Also, avoid the messy close. You can close early, and effectively. When closing, end with a clear *thank you*, then invite applause. Don't drift into "So, any questions?" No one knows whether to clap. And, a silent auditorium drags down the energy at the crucial conclusion of your talk.

10. Ended Late. Running over time ruins a great talk. Show your contact info or social media link on screen. Manage the question/answer time. Offer a closing summary at the very end. (I like to end with a ukulele parody song, but that's just me.)

> **The Lesson Learned**
>
> Use your body and voice to intrigue. Why does the topic grab you? Use strong visuals, Tell a great story. Use these 10 tips to improve your next talk.

(For specific examples of good and bad slides, see the blog posts linked in **Appendix 1.2**.)

APSO Notes, Maxwell, and the Meeting Before the Meeting

My personal pediatric medical record (below) is a testament to the efficiency of handwritten doctors' notes. My pediatrician was able to fit 13 years of my life on a single sheet of paper.

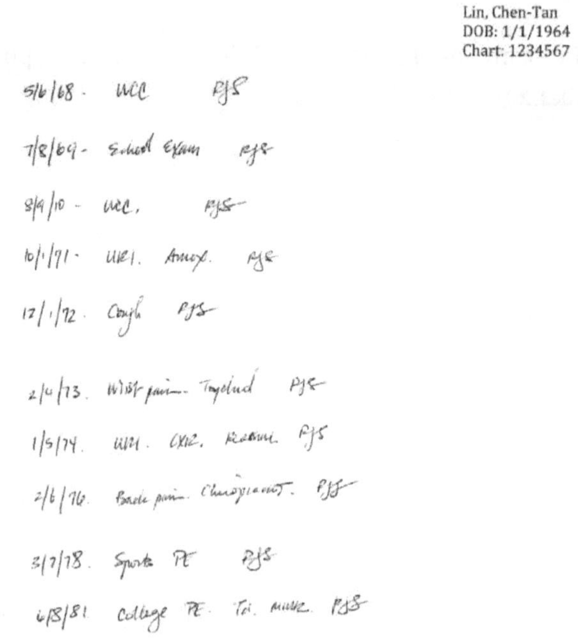

Lin, Chen-Tan
DOB: 1/1/1964
Chart: 1234567

5/6/68. WCC RJS
7/8/69. School Exam RJS
8/9/70. WCC. RJS
10/1/71. URI. Amox. RJS
12/1/72. Cough RJS
2/4/73. Wrist pain. Tylenol RJS
1/5/74. URI. CXR. Recheck. RJS
2/6/76. Back pain. Chiropractor. RJS
3/7/78. Sports PE RJS
6/8/81 College PE. Td. MMR. RJS

WCC—Well-child check, URI—upper respiratory infection, Amox—amoxicillin, CXR—chest X-ray, PE—physical examination, Td—tetanus-diphtheria vaccine, MMR—measles, mumps & rubella. "RJS" is my doctor's initials.

This approach certainly saved my doctor time after each visit. But that efficiency came at the expense of readability. Look at that chicken scratch! The arrival of the electronic health record and of digital notetaking was certain to make everything nice and legible.

In 2009, we were eight long years into deploying the Allscripts TouchWorks EHR. We were making the messy transition from handwritten notes to typed ones. Sure enough, typed digital notes were easier to read and faster to find than waiting for a courier to bring a paper chart full of illegible scribbles. But electronic notes came with their own problems.

For one, it was now possible to copy and paste parts of previous notes and test results. The ease of "reusing" information meant that notes got longer. Well-intended federal regulation made things even worse. These notes were required to include statements like, "I certify that we screened this patient for smoking. We screened this patient for depression. We ensured the medication list was reconciled. The patient noted they were not worried about falling down. I spent more than 50% of our 40 minutes together in counseling and arranging care." And so on.

If progress notes are for doctors to see what other doctors were thinking, these bloated notes obstructed the view. We had exchanged illegible scribble for unreadable scroll. It was harder than ever to find the information we actually needed.

Progress notes were traditionally written in a format called SOAP: Subjective (the description of the patient's history and symptoms), Objective (physical examination or test result findings), Assessment (clinical reasoning and diagnosis) and Plan (what treatment or actions to do next). It has been this way since Dr. Larry Weed proposed this organized way of tracking patient data in the 1960s.

Back when doctors wrote brief notes by hand, SOAP notes were terrific. Even today, a surgeon's SOAP note can be brief and useful:

S: Post-op day 2. Right inguinal hernia repair. Pain improving. Not passing gas yet.

O: Afebrile, vitals stable. R groin wound is clean, dry, no redness. Ambulates well.

A/P: Hernia repair healing well. Teach dressing change. Home in AM if passing gas.

In contrast, our EHR notes in 2009 were full of regulatory language, audit trail signatures, and time stamps that could stretch to 15 pages for a single visit. It was like we were saying, "Look at all the work I'm doing! I'm meeting all the regulatory and quality benchmarks!" And yet, these long notes no longer helped doctors communicate their thinking.

Here is an example of a terrible emergency department note, circa 2010. Notice the circled clinically important elements in this 15-page note. The last page is all signatures and time stamps. The all-important Assessment and Plan hides on page 11.

What is worse, these long notes are particularly unpleasant to read on a screen. Even when there are hundreds of progress notes on a patient, doctors were unlikely to read more than two or three before they stop looking. Every doctor wants to understand the course of the patient's illness and what previous clinicians thought about the patient: That is the Subjective/HPI section, and then the Assessment and Plan. The remainder, although important for billing and regulatory compliance, is just noise. There is never enough time to sift through all the data, so we read the most recent note or two and hope for the best.

I knew that digital notes in the EHR were the future. But we were drowning in data, with a shortage of clarity. What could we do to improve this sad state of affairs?

One major problem was scrolling. The pertinent information in notes was always buried, and doctors had to scroll through pages to get to it. My solution: APSO notes! Invert the format, so that the A/P is the first thing you see when you open a progress note. A finished progress note should show: Assessment/Plan, then Subjective, then Objective, and finally all the regulatory language at the end, where it is present for auditors and non-intrusive for clinicians.

Yes, the long-term, better solution is to attack so-called "note bloat" by training doctors to write shorter notes, and simultaneously reduce the length of regulatory and billing statements in our templates. Those solutions will take years of education and culture change. In the meantime, how hard would it be to teach docs to just put A/P at the top of the note?

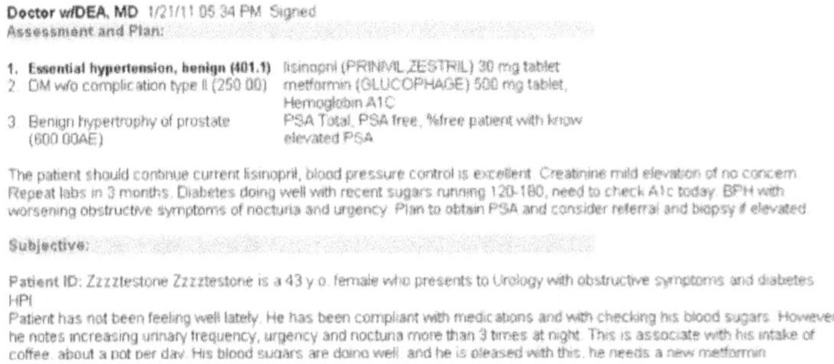
Doctor w/DEA, MD 1/21/11 05 34 PM Signed
Assessment and Plan:

1. **Essential hypertension, benign (401.1)**	lisinopril (PRINIVIL,ZESTRIL) 30 mg tablet
2. DM w/o complication type II (250 00)	metformin (GLUCOPHAGE) 500 mg tablet, Hemoglobin A1C
3. Benign hypertrophy of prostate (600 00AE)	PSA Total, PSA free, %free patient with know elevated PSA

The patient should continue current lisinopril, blood pressure control is excellent Creatinine mild elevation of no concern Repeat labs in 3 months. Diabetes doing well with recent sugars running 120-180, need to check A1c today BPH with worsening obstructive symptoms of nocturia and urgency Plan to obtain PSA and consider referral and biopsy if elevated.

Subjective:

Patient ID: Zzzztestone Zzzztestone is a 43 y o. female who presents to Urology with obstructive symptoms and diabetes
HPI
Patient has not been feeling well lately. He has been compliant with medications and with checking his blood sugars. However, he notes increasing urinary frequency, urgency and noctuna more than 3 times at night. This is associate with his intake of coffee. about a pot per day. His blood sugars are doing well and he is pleased with this. he needs a new metformin

With APSO, a physician would still write their note in the same sequence: skip down and write the Subjective, Objective, and then move back to the top and complete the Assessment and Plan. It's the simplest change possible—same SOAP sequence, but the final report is APSO for easier reading.

If all progress notes are written in APSO, then reviewing a complex patient chart is dramatically easier. Open the first note—A/P is visible without scrolling. Open the second note—A/P visible without scrolling. It is much faster to browse and get a concentrated dose of clinical thinking across many notes.

When I tried it with my own notes, I thought to myself: "This is the way." All that remained was to get my colleagues on board. How hard could it be to change doctor behavior? If you've read this far, you already know the answer.

I built a new APSO note template in the EHR, an alternative to the standard SOAP. I user-tested it with about 80 colleagues, who all agreed, "This does not take longer and is much easier to read." I launched the APSO note template and held my breath.

The result: APSO was adopted by 70%... of my close friends. Womp womp. I was hoping for over 90%.

On the one hand, I was pleased—70% meant a majority of clinicians were using the format. However, thinking about EHR usability, I realized that two thirds adoption was not an improvement. With two thirds APSO and one third SOAP notes in a chart, clinicians would have to search both the top *and* the bottom of each note to find the A/P. We needed universal adoption to fully realize the benefits.

In other words, I needed leadership consensus to make a universal change.

This is where my past caught up with me. Years before, I had tried to convince medical leaders at my organization to adopt OpenNotes and allow patients to read their doctors' notes about their visits. I had walked into the big meeting and presented my best scientific evidence in favor of OpenNotes. My attempt to change minds in the big meeting failed miserably (see **The Naïve Scientist Part 2: "But, the P-Value!"** on page 193). I knew my idea was good. I did *not* know about resistance to change.

Once again, with APSO notes, I knew I had a great idea. I knew I would face resistance. I was determined not to repeat my failure.

In the years since that failure, I had read John C. Maxwell's book *Leadership Gold*. To paraphrase Maxwell: "If you go into a meeting and you do not already know how everyone will vote, you have not done your job."

The multidisciplinary Ambulatory Medical Directors group has 40 members. They are a rowdy, tough-to-wrangle group with strong egos and lots of differing opinions. For such an important decision, I decided to meet one-on-one with each influential member of the group. Maxwell calls this "the meeting before the meeting."

CT: "Don't you hate how hard it is to read the notes of [other specialty]?"

Medical Director: "Yes! Their notes are terrible."

CT: "I agree. I'm going to require them to switch to APSO format so their notes will be easier to read."

MD: "That sounds great. They really need to change."

CT: "However, if I'm going to get them to change to APSO, I need your specialty to change to APSO. We need to be unified and apply pressure to them."

MD: "But our notes are fine. We don't need to change."

CT: "I know! Your notes are fine! However, I can't do this without you. The actual process change for you is minimal. If you agree, I will switch the template for your specialty. For a minimal sacrifice on your part, we all benefit from *their* notes being more readable. Can I count on you to get your team to switch?"

MD: "Hmm. I suppose it is worth it. You're sure you can get them to change, though?"

CT: "Yup, you give me your word that the docs in your specialty will use APSO format, and I will get them to change."

MD: "Deal. Makes sense. Minimal change for us, big improvement on their part."

I repeated this same conversation with every medical director, each of whom had the same thing to say: "We're great, but those guys are terrible."

As a result, the big meeting went smoothly:

"CT is proposing APSO format as a standard for all progress notes written in the EHR, in all specialties, inpatient, emergency department, and outpatient. Any discussion or objections?"

"Nope. Makes sense to me."

"Motion approved."

I had invested 20 hours of planning and individual meetings for a five-minute agenda item.

It may seem impractical to have a meeting before every meeting. Yes, it is. I pull out this tactic only for the big decisions. To this day, over a decade later, primary care notes are 100% APSO format across a million clinic visits every year. Specialties notes are 88% APSO.

> **The Lesson Learned**
>
> A big meeting is not the place to allow those resistant to change to gather strength from others. Conflict and negotiations should take place beforehand. The meeting before the meeting is a useful tool in the Psycho 80 toolbox.

2026 Addendum—Ambient notes have arrived in our organization, with generative AI tools that listen to the patient–physician conversation and write a completed progress note in less than a minute. In our organization, we still configured the output of the ambient note in APSO format and continue to find this arrangement useful.

Sprint Informatics

EHRs and physicians are not natural friends. One of the most difficult but most rewarding informatics projects we ever undertook was building an EHR Sprint Optimization Program to help physicians, APPs, and their teams. From the very outset, the Sprint concept faced enormous challenges. With some clever choices, some memorable stories and quotes, and some lucky breaks, we built a trustworthy team and grew a national reputation. Here's how it happened.

"That Epic Thing You Wrote? It Sucks."

A colleague said this to me one day in 2016. They evidently misunderstood that, no, I did not personally write the million lines of code that make up our electronic health record. My first thought: "That is hilarious, that my colleague thinks the EHR is something built by one person." Then I had a sobering thought—probably a fair number of colleagues would agree that the EHR "sucks."

EHR "suckiness" is not entirely fair. The EHR is essential for any health system to survive in a data-driven future. Unlike rooms full of paper charts, an EHR can spot patterns across millions of patients and gently nudge physicians/APPs to prescribe aspirin after heart attacks, statin drugs for diabetes patients, and spot hidden depression symptoms when neither patients nor their doctors realize it.

EHRs also connect patient data across the country and connect doctors' recommendations with patients and their families. The friction of delivering paper charts, or faxing reams of paper, or scribbling some notes for patients to take home—those days are mostly gone.

And yet, complicated tools can be hard to use. Some colleagues were leaving medicine, burned out from the burdens of using the EHR. Clinicians went into medicine to see and treat patients. In decades past, physicians saw the paper medical record as an afterthought: a record of their actions required for continuity of care and also to be paid for the work. Scribble some words and you're done.

Now, the EHR is more powerful and more intrusive. Throughout a full day of seeing clinic and/or hospital patients, a clinician has many EHR tasks: Write notes on every patient. Place all the prescription and test orders. Respond to all the reminders and alerts about their patients. Follow all the regulatory and legal requirements. Respond to phone calls and online messages from nurses and patients and colleagues. Read and comprehend the exponentially growing set of test results and data available for each patient. And on it goes. Is it any wonder that some feel frustrated with the EHR?

How might we help physicians use the EHR effectively? If we worked in a small clinic, one could take a few hours, walk among colleagues, gather feedback, and do some at-the-elbow coaching to help teach efficient use of the EHR.

In contrast, large organizations like ours have thousands of clinicians and hundreds of clinic locations. Hiring and training enough people to support busy clinicians and staff is difficult.

In 2016, five years into our Epic EHR journey, we had tried many ways to inform and upskill our physicians and APPs. However, our email newsletters, leadership meeting announcements, lunch-and-learn sessions and EHR "rounders" (trainers who would walk around between-clinics and within-hospitals) had not been successful.

Challenge #1—Newsletters, Training Sessions, and Rounders

We knew that physician EHR skill was a bell curve. There were a few highly proficient EHR users and a few were vocal, dissatisfied users, but most were among the "silent non-complaining majority." Yet, this majority of users were still not using the EHR efficiently. Many colleagues still spent "pajama time" in the evenings catching up on unfinished notes or tasks left over from the day.

"Perhaps a monthly newsletter might help!" we thought. We could email everyone with tips and tricks for using the EHR. "If we make it fun, surely everyone will read it!" Incorrect. We learned that almost no one reads newsletters.

"Fine. If people won't read our emails, then we'll have to teach them in person." We set up in-person and virtual training sessions—no one came. We set up a telephone help desk—the response time was too slow for busy docs. We sent skilled EHR trainers to clinics, but clinicians were too busy to stop and learn something new.

We were stumped. How might we reach our burned-out colleagues?

Challenge #2—Specialty Requests are Denied

There was a second problem. With the limited budget for optimization and a constant stream of fix requests, we had difficulty keeping up. Our overworked EHR analyst team were focused on making general improvements that helped all physicians. It was never their priority to do special work to benefit a select few.

By default, then, specialty requests like blood sugar templates for the diabetes clinic, or auto-tracking rejection drugs for the heart transplant service, or post-operative mobility scores for orthopedics, never got built.

We needed a way to reduce the EHR burden, improve specialty tools, and just make the EHR more useful. How might we do this? Where do we start?

Two ideas came from outside of our organization.

Big Idea #1—Apple's Genius Bar

Those of us who experienced this service to fix an iPhone or Mac loved the personal service. We had previously stolen the Genius Bar idea (see **Genius and Ninja Stickers** on page 88) and found success with docs by going into clinics, observing, and teaching at-the-elbow to improve their EHR skills.

Big Idea #2—Integrated Architects

Dr. David Butler, then CMIO at Sutter Health in California, came up with the idea of "integrated architects," a small team of energetic 20-year-olds with complementary skills: a web designer, a report writer, an EHR analyst, etc.). He gave the team a simple directive: "Go into the clinic. You have two weeks. I don't care what you do, just make it better." It was a brilliant, inexpensive, creative solution. The team's deep immersion and their broad skillsets allowed them to build rapid protypes and deploy elegant, simple solutions.

I heard Dr. Butler speak at a conference and thought, "*This* is the way to disrupt the old model."

The old model: Put physicians in an eight-hour intro class (they will only absorb a small fraction of what is taught). Then, hope they get better as they use the EHR with patients (they're too busy to learn new skills).

Send newsletters and hope they set aside time to read and learn. (They won't).

The new model: Instead of sending newsletters, conduct intensive immersions with a small, skilled group. Build tools, reinforce efficient workflows, teach new skills. And *voilà*! EHR satisfaction improves!

I was excited to bring this idea home and get started right away. I put together a proposal and prepared for my moment of glory.

What happened?

Nothing.

The requested proposal was denied.

I had written up budget options for an EHR optimization team: a small team costing $200,000, a medium team costing $500,000, or a large, 11-person team costing about $1 million. After months of negotiating, the answer was: None of these. Too expensive. After implementing an expensive EHR, leaders found it difficult to understand the need for funding optimization training.

We had a small source of hope: The CIO saw that the EHR efficiency scores of individual docs were very different, even within a single clinic. A very efficient doc worked side-by-side with a doc who struggled to use the EHR. In other words, the CIO realized that even physicians who worked next to each other didn't naturally share EHR skills by diffusion.

A physician with poor EHR use is like a driver buying a sports car and not learning how to shift out of first gear. Optimization training would have to get to individual docs. We would have to reach every doctor and put them through an intensive driving school.

The CIO began to support the idea of local optimization and intensive training. Even with CIO support, however, the budget was tight and the EHR optimization team did not make the cut. Over the next two years, I waited, improved the proposal, and waited some more.

In my third year of waiting, I decided—enough is enough. I set aside several days of time over a two-week period and asked a couple members of my physician informatics team to do the same. I begged the training director and ambulatory analyst director to lend me a couple of their team members. "Just a few days over two weeks to try something new, then they can go back to their regular jobs." I called in some favors. Amazingly, I managed to gather six people to do the work.

The Power Rangers were assembled! But which clinic should we choose to do our first intensive optimization? We needed an underperforming clinic, but we also didn't want to go into the lion's den with the most unhappy physicians. We chose a middle-tier clinic with an effective leader, a cohesive clinic team, and average-efficiency docs.

These clinicians were driving in first gear and generally heading in the right direction, but we could pave the road and teach them to shift to a higher gear. For us, endocrinology ended up being a good choice.

We set a go-live date. I got so excited launching our secret project that I forgot to announce the specific date to several key people: that is, the clinic manager and the entire clinic we were going to optimize. As a result, we showed up on a Monday morning to a group of puzzled faces asking, "What are you doing here? What is this about?"

After some initial confusion, clinic leaders let us "borrow" their conference room in the middle of the clinic and set up shop: six laptops, some wall-sized sticky notes with big markers, lots of yellow pads. Once settled, we wandered the hallways looking for unsuspecting doctors and nurses to shadow. We would look over their shoulder and ask annoying questions like, "What are you trying to do there, on the computer?" or "What are you trying to look up?" Then we would point out a shortcut here, a quick-action there.

Word-of-mouth spread, and by day three we were very popular: "Hey guys, go into the conference room and talk to the Epic team: they're here to help us!" and "You guys are like a secret organization of hackers, you come in and make things better."

Modeling ourselves after the "agile sprint" approach that is common in software engineering, we would check in with the clinic leaders every day, agree on the biggest problems, and tackle them with our full attention. We built a new diabetes flowsheet that summarized a patient's response to medication. We built a checklist that compiled useful information at a glance for thyroid nodule tracking.

Rapid improvements were surprisingly rewarding for analysts who rarely saw the smiles of those using the new tool. Rapid-cycle work was also rewarding for the physicians. One told us, "I can see things getting better. It feels like someone cares."

Surprisingly, for all the improvements and new tools that our analysts built for the endocrine clinic, 70% of the benefit came from training physicians and staff on existing tools, with only 30% from building new tools or fixing broken ones.

Observing over a physician's shoulder as they laboriously copied sections of their previous note: "Next time try using that button." "What does that do?" "It eliminates about 17 clicks that you just did." "Oh my god. You're kidding me."

We were walking on air at the end of this EHR Sprint with our "stolen" team. We knew we had done something big. We had reached almost every doctor in the clinic with multiple one-on-one training sessions. We had built several long-awaited tracking tools. We had heard, "This is one of the best things our organization has ever done for doctors."

But then—our borrowed time was up. My "stolen" team was to disband and get back to our regular jobs. I would have to figure out how to get funded for a second Sprint. How could I keep this from being another promising "pilot project" that fizzled and disappeared? That was my next challenge.

The Lesson Learned

Sometimes you have to steal a team and then ask forgiveness. Find a way to start a project for free. Show them what you can do. And, surprisingly, EHR optimization is 70% training on existing tools and only 30% "building or fixing stuff."

"Put Those People Back Where You Found Them!"

When our endocrinology Sprint had concluded, our team disbanded. I was a blur of emotions: among them excitement at our success and anxiety at facing the funding struggle again. As I was summarizing the results of our pilot, I heard some surprising news.

The chairman of endocrinology had bragged at the department chairs meeting about his Sprint optimization experience. Since EHR dissatisfaction was common among physicians, there was great clamor among the chairs of other departments. "What is a 'Sprint?' Why didn't we get one? Who do I talk to?"

Soon, there were many requests for an EHR Sprint in other departments. Over the coming weeks, the CEO and the CIO allowed us to borrow the team for a second and then a third Sprint. We racked up additional successes.

After my years of unsuccessful requests to fund an EHR optimization team—a concept that no one had ever heard of—I finally got support by stealing a team and demonstrating what it could do.

To help spread the word, I put together a one-page PDF summary (see **The Superpower That Is the One-Pager** on page 83) about the EHR Sprint. In it, I explained the idea of a Sprint and showed hard data: improvements in teamwork and reduced pajama time. I also cited heartfelt quotes from physicians: "You made me a better mother. I get home in time for dinner with my family." And "I no longer contemplate retirement to escape the EHR."

To show further evidence, we looked for an accepted measure for user satisfaction. At the time, there was no healthcare industry satisfaction metric for EHR use, so we stole one from the computer industry: the NPS, or Net Promoter Score.[1]

The NPS is determined by asking all participants the following question: "How likely is it that you would recommend (XYZ service or product) to a friend or colleague?" The possible answers range from zero, meaning "not at all," to 10 or "would absolutely recommend." Replies of 9 or 10 are considered "Promoters." They are happy about this project and may tell others. Replies of 7 or 8 are thrown out as "neutral" without strong opinion. Replies of 6 and below are considered "Detractors." I think of these sixes as "Minnesota-nice" for "I don't like this." Take the percentage of "promoters" and subtract the percentage of "detractors," and you get a single number that balances promoters, neutral responses, and detractors. This is the Net Promoter Score.

The best possible NPS would be 100, in which everyone is an avid promoter. The worst would be negative 100, meaning everyone is a detractor. When promoters and detractors cancel out, you get a zero—which means "meh." A so-so rating.

When we gathered metrics on EHR use, our NPS was negative 44 before the Sprint, and zero at the end. In other words, we went from "I hate this" to "It's okay." That may seem undesirable, but a 44-point improvement in two weeks is remarkable. Also, responses to "Would you recommend Sprint to others" showed a whopping NPS of 70. Most respondents rated the Sprint team as a 9 or 10. Those numbers are comparable to Apple Computer, a standout for loyal customer relationships in the computer industry.[2]

Having distributed the one-pager, we received word that our executives wanted Sprint to continue. But funding was still tight.

To get around this, our CIO's approach was to permanently reallocate existing IT staff to the Sprint. Instead of seeking funding for a new team, we explained to clinical and operational leaders that some existing IT services would slow down. Then we stole trainers and analysts from other teams to create a permanent Sprint team. Our clinical leaders agreed that the tradeoff was worthwhile, and we were allowed to keep "that team you stole!"

It was a joyful time. Over the coming weeks, we built our dream team out of existing IT members: a physician informaticist leader, a nurse informaticist, a project manager, four EHR trainers, and four IT analysts.

The good news: every clinic we Sprinted was an unqualified success. Recipients were grateful, physicians and teams were more efficient, and we completed a growing number of specialty-specific tools and workflows. There was better patient care with less burnout among staff.

Even then, it was not without challenges.

How would we help differently sized clinics? How might we allocate precious Sprints between extra-large academic clinics of 90 neurologists and tiny primary care practices of one or two docs in rural areas? Solution: Break large clinics into separate Sprints, with about 30 providers and their immediate support staff per group. For small clinics, group them geographically and Sprint them together in groups of about 30 providers.

How do we balance speed versus consensus? How might we balance the need for nimble, single-clinic optimizations against building consensus across similar specialty clinicians throughout our organization? Solution: Every time we Sprint a specialty clinic, we call up leaders of same-specialty clinics for online consensus-building meetings.

How long should a Sprint improvement session last? We chose two weeks—long enough to meet every physician/APP for multiple training sessions and build specialty tools, but short enough to create a sense of urgency.

How would we decide what was "in scope"? Some physician requests were to "fix my keyboard tray" or "get my printer working." We could have said "not my problem." Instead, we knew all the right support teams, and we took those opportunities to place trouble tickets. Because we skipped nothing, we developed a reputation of being general problem-solvers.

How did we keep our Sprint team from burning out? Sprint members were supporting clinics from 8:00 a.m. until well into the evening, day after day. Solution: Intersperse rest weeks so they can catch up on work from prior Sprints, take breaks, pursue mini-projects, and serve as a resource for other teams.

How might we protect the Sprint team from becoming the "I know a guy" to thousands of people? A growing number of clinicians learned the emails and phone numbers of Sprint members and forevermore considered them a personal, on-call source of help. Solution: Build a disciplined response system to reroute such requests to the help desk.

How might we be efficient from day one of the Sprint? Learning from the disorganization of our first Sprint, we began meeting with clinics 90 days before the go-live date. We would meet and ask questions. Three months ahead was the right time to block out hour-long slots in doctors' clinic schedules so they would have time for one-on-one training without having to reschedule patients. Then we would plan 60- and 30-day check-ins with clinic leaders to identify and prioritize our work together: Who struggled with the EHR? Which workflows were particularly troublesome? Prescription refills? Specialty referrals? In Basket messages? By the time we reached the first actual day of Sprint, we had done our homework and could hit the ground running.

How might Sprint help staff members, not just physicians? In our first Sprint,[3] we had focused solely on physicians, the most burned-out members of the team. Yes, the Sprint helped physicians become more efficient, but as a result, their medical assistants and nurses many more tasks than before. The unexpected consequence was a skyrocketing workload and burnout for the MAs and RNs! Solution: We realized that Sprint would need to optimize the work of not just physicians but the rest of the clinic team: MA, RN, front desk. Teaching individual skills was not enough. We should teach teamwork. Sometimes the right solution is to stop the ping-pong EHR messages, sit down together, and problem-solve in person.

In the service of helping physicians learn teamwork and efficiency, these were wonderful problems to have, and we enjoyed solving them.

The Lesson Learned

You stole a team. Now what? Try new stuff. Use the PDSA (plan-do-study-act) cycle. Have your beneficiaries brag for you. Keep building momentum.

Sprint: Setbacks and Evolution

Following its uncertain start in 2017, Sprint became an effective service. After a Sprint, physicians saved 20 minutes per provider, per day.[1] Word spread, and clinics began to look forward to our arrival. We scored well on external measures of "EHR burden vs. efficiency" in the Stanford Wellness survey and received the American Medical Association Joy in Medicine Gold Award, and others.

We developed a schedule of 17 Sprints a year—two weeks on Sprint, one week off. These intermediary weeks were crucial for team sanity.

Then, in March of 2020, the COVID-19 pandemic arrived in Colorado. In-person Sprints came to a screeching halt. Until then, we had thought the value of the Sprint was our in-person immersion in the clinic. With close observation and daily huddles, we built trusting relationships and strong collaboration. Then, as the virus arrived, clinics began to shut down. And so did we.

Thankfully, we were not idle for long. Under federal emergency authorization, telehealth visits began to be covered by medical insurance, and there was an immediate need to teach thousands of physicians and their teams how to use video conferencing tools to see patients.

The Sprint teams sprang into action. In less than two weeks, they had taught thousands of colleagues over Zoom conference meetings, how to use telehealth tools integrated into the EHR to schedule and see patients. Our organization went from seven telehealth visits a day to over 1,000.[2]

Following the successful launch of telehealth in every clinic, and the ongoing stay-at-home order, the Sprint team turned to reinventing itself as a virtual-only service. We feared Sprints would only be effective when we could stroll the halls and offer literal at-the-elbow training to physicians. We worried that clinics would ignore us or that users wouldn't learn as well. Gratifyingly, we were wrong. Not only were our Sprints successful, the NPS scores actually increased.

This forced move to virtual-only Sprints showed us the benefits of collapsing the physical distance. Virtual sessions allowed us to group together clinics that were geographically distant but had the same specialty—specialists from several states who had never met each other began to collaborate. We eliminated unnecessary variation in our order preference lists, our note templates, our synopsis reports, and our order sets. Virtual training also allowed physicians who had family responsibilities, administrative jobs, or other geographic barriers to attend. Thanks to the flexibility of telehealth, Sprint participation increased.[3]

As our organization continued to add hospitals and clinics, Sprint trainers helped with onboarding and support of physicians and staff at new locations. Though not informaticists by title, the trainers deeply understood physician workflows. After a year of intensive Sprint teamwork and experience supporting busy practices, they functioned like physician informaticists.

Sprint continues to evolve. We are adding short-form (that is, TikTok-length) educational videos (see **What TikTok Teaches Us About Communication** on page 75). We have added self-guided uPerform and Amplifire modules to the Sprint curriculum. We cross-train our newer EHR educators with our Sprint team. As generative AI tools mature, we look to Sprint's next evolution.

> **The Lesson Learned**
>
> An effective optimization team should constantly update its teaching topics and methods based on feedback to stay relevant and useful. Such teams can respond to unexpected crises.

(For additional studies of EHR Sprints, their different iterations, and their effectiveness, see **Appendix 1**.)

INTRIGUE—All the Lessons Learned

Intro—If You Build It, They Will Not Come

Workflow eats technology for lunch. Even great technical achievements stumble without INTRIGUE, getting clinicians interested, and adjusting workflow. Think like a user: WIIFM ("what's in it for me?"). Tech alone is not enough.

What Is a Sacred Bundle, and Why Do You Need One?

Stories make up the DNA of your organization. Your team should know them. Collect and tell the stories that INTRIGUE your team and build your culture. What objects go in your sacred bundle? What stories do you tell?

A Surgeon's Friend

People are constantly telling you how well you communicate. Listen hard. If you're not getting through, it's on you. A simple message carries more impact. The responsibility of communicating lies with the sender.

What TikTok Teaches Us About Communication

Attention is short. Videos should teach, entertain, and be no more than a minute or two. As agents of change, and educators, how will we use this?

Paternalism and the Patient Portal

Unwind paternalism toward patients and toward EHR users. A patient-centric attitude makes a better doctor. A user-centric attitude improves informatics.

Haiku Poetry

A few syllables / Attention, whimsy, meaning / Intrigue with poems.

The Superpower That Is the One-Pager

One page is the perfect size to INTRIGUE your boss. Is it a report? Include data and stories. Make it brief and clear. Is it a request? Define the problem, have a clear ask.

What is Your Boss's Learning Style?

Communicating with patients, colleagues and direct reports is important. It is even more crucial to communicate effectively with your boss.

Genius and Ninja Stickers

"Let me teach you something!" is less engaging than "Are you a genius?" And "I'm too busy" becomes "Thanks for bringing joy to our work." INTRIGUE your colleagues. Fun and effective can be cheap and easy.

The DOs and DON'Ts of an Effective Presentation

Use your body and voice to intrigue. Why does the topic grab *you*? Use strong visuals, Tell a great story. Use these 10 tips to improve your next talk.

APSO Notes, Maxwell, and the Meeting Before the Meeting

A big meeting is *not* the place to allow those resistant to change to gather strength from others. Conflict and negotiations should take place beforehand. The meeting before the meeting is a useful tool in the Psycho 80 toolbox.

"That Epic Thing You Wrote? It Sucks."

Sometimes you have to steal a team, then ask forgiveness. Find a way to start a project for free. Show them what you can do.

"Put Those People Back Where You Found Them!"

You stole a team. Now what? Try new stuff. Use the PDSA (plan-do-study-act) cycle. Have your beneficiaries brag for you. Keep building momentum.

Sprint: Setbacks and Evolution

An effective optimization team should constantly update its teaching topics and methods based on feedback to stay relevant and useful. Such teams can respond to unexpected crises.

INFLUENCE

During your career, you will make or influence decisions. Some decisions will be small and affect only a few people; others will be big and scary and affect many people. How much information do you need for each? How quickly must you decide? Must you limit yourself to the options presented? Once you've decided, what then?

Even if you aren't the one making difficult decisions, informaticists often support leaders who do. If you are not the decision-maker, how can you help your leader choose wisely?

Intro—Is It Time to Be the Boss?

In the mid-2000s, I was working as a Senior Medical Director of Informatics when my CIO boss announced his retirement. He met with me to discuss my future. "Have you thought about being a CIO?"

What made our two jobs different was their scope. A CMIO focuses on the IT and electronic health record technology that affects physicians and APPs, whereas a CIO's job is much broader: wired and wireless networks, telephones, pagers, server farms running applications, many applications and databases themselves, plus the relationships with vendors who supply the software and hardware—not to mention the privacy and security concerns that come with this infrastructure. In that moment, it seemed like a lot for me.

As I expressed my doubts about my potential as a CIO, he suggested: "Why don't we send you to the CIO bootcamp? There, you can meet with others who are considering the CIO job, spend a few days with instructors who are themselves experienced CIOs, and try the role on for yourself." Who would say no to that?

The College of Health Information Management Executives (CHIME) is an organization comprising about 1000 healthcare CIOs. Twice a year, they run a bootcamp for aspiring CIOs. Along with about 30 other attendees, I spent three days in lectures and group exercises simulating CIO challenges. We dove into seven-person working groups to solve leadership problems.

For example, this scenario: How might we combine the IT staffs of two merging hospitals? One hospital has an EHR with a terrible technology reputation but is staffed by an outstanding, cohesive IT team that has successfully deployed their EHR. Another hospital has a best-in-class EHR but an IT support crew and a medical staff with very low morale.

Or: Here are 18 proposed projects for next year. We can only fund about half of them. Some projects depend on others, some are high-profile projects for key individuals, others are non-sexy crucial infrastructure improvements. How do you build consensus and make good decisions?

Or this one: How might we build a financial and clinical case for purchasing a new technology in an organization of 50 clinics? Clinicians love the idea, but administrative leaders don't support the large financial cost.

I really enjoyed the bootcamp. On my return to Denver, I reflected that leaders in information technology faced similar challenges, whether as a CIO or CMIO. The fragility of human relationships, the need for clear thinking, the application of leadership skills—the core aspects were the same.

I realized I *could* do the CIO job. The content differed: CIOs are more focused on technologies in a hospital whereas CMIOs sat more at the intersection of clinical medicine, teamwork and technology. But, the skills and the willingness to make decisions in ambiguous settings, was the same.

This section covers different kinds of INFLUENCE and when to use them. Whether you are the boss, whether you lead by consensus or by decree, there are some common threads we will explore.

The Lesson Learned

Much of leadership is about building teams, making good decisions, and carrying them out. Whether you are a CIO, a CMIO, or some other IT or informatics leader, good leaders have common skills and characteristics.

On Machiavellian Leadership

In 1997, I was a freshly hired assistant professor at an academic medical center, and I knew nothing about leadership. I had been offered a part-time position as "physician liaison," but I was uncertain whether I had the skills to succeed. Then I met my mentor.

John Sbarbaro was then president of University Physicians, Inc (UPI), the organization that managed finances for the University of Colorado School of Medicine physicians. I knew Dr. Sbarbaro from his presidency and his occasional lectures at our clinical faculty meetings. He was clearly a respected leader with a fair number of accomplishments to his name. Seeking guidance and direction, I requested an appointment to meet with him and asked his opinion of this opportunity for me.

"Well," he said, "You should read the operating manual for academic medical centers."

"*What*!" I said. "There's an operating manual for academic medical centers? What is it called?" I whipped out my pen and paper, ready to scribble down my next book purchase.

"I have a copy right here." He turned to his bookshelf, took down a small book, and handed it to me. "It was written in the sixteenth century by Niccolò Machiavelli. It's called *The Prince*."

I was shocked. How could a text from 500 years ago be of use to me in modern society? And wasn't Machiavelli that *power-hungry backstabbing* political operative? I wasn't eager to be known as a slick operator.

And then I read it. And Dr. Sbarbaro was exactly right.

The Prince taught me a set of principles for how to approach and manage change. Machiavelli (and Dr. Sbarbaro) understood that change is necessary for progress and improvement. Unfortunately, no one likes change. People are resistant to change, and they take their frustration out on the person leading it. Here's what *The Prince* recommends in response.

Machiavelli on the Inevitability of Resistance

> *For the introducer has all those who benefit from the old orders as enemies, and he has lukewarm defenders in all those who might benefit from the new orders...*
>
> —*The Prince*, Chapter VI

How do you push through when a large project gets stalled?

Large organizations are run by committees. They are designed for standardized work and reliability. They value stability over speed. President George HW Bush used to speak about helpful Americans being "a thousand points of light." Well, committees in large organizations are "a thousand points of veto." When it comes to new ideas, the default position is generally "Sorry, no."

Of course, these committees are important. Without them, undisciplined decisions would cause conflict and chaos. On the other hand, these bureaucracies stymie the individual innovator. Knowing this, how might an individual propose an innovative idea and help it survive in this complex environment?

If you think of yourself as insightful or innovative, you're probably used to having good ideas. Unfortunately, it is very hard to bring those ideas to life. To do so you must:

- Speak up about your idea *and* get someone to listen.
- Prove that it is better *and* that it aligns with organization's goals.
- Gather support *and* convince colleagues your idea should be prioritized.
- Find the tools or tech to implement the idea *and* teach others to use it.
- Suffer setbacks and overcome barriers *and* commit to seeing it through.

Consider these steps in the context of your latest initiative. Did you get stuck? If so, where?

It is extremely rare for a big project to get off the ground immediately and then proceed smoothly to a successful design, deployment, and widespread adoption. But even within that challenge, some organizations do a better job at encouraging innovation. Hawaii Pacific Health in Honolulu is known for their program "Getting Rid of Stupid Stuff" or GROSS.[1] Having a leader-supported program to generate new ideas encourages employees to speak up. It rewards employees by capturing their ideas, removing obstacles, and bringing the ideas to life.

In the meantime, most of us do not work in rare organizations like this. Therefore, innovators who wish to succeed must cultivate grit and persistence. Nothing worth doing is easy.

For instance: make a list of Big Hairy Audacious Goals, or "BHAGs," to hang in your office (see **Getting Things Done in an Age of Acceleration** on page 275). These are reminders of what you may someday accomplish if you stick with it. If you run into passionate opponents, or strong defenders of the status quo, you may have to back down and concede the battle for today and consider when/how to try again.

The lesson taught in *The Prince* is that resistance is inevitable. It doesn't matter how good your good idea is. Instead of getting discouraged, figure out how you can INFLUENCE those who might stand in your way.

For example, you could...

...make sure you're **not missing something** about the situation (see **The Sonar Sensor Conspiracy** on page 56).

...distill your idea to a **one-page explanation** (see **The Superpower That Is the One-Pager** on page 83).

...**gather individual consensus** from leaders (see **APSO Notes, Maxwell, and the Meeting Before the Meeting** on page 98).

...force through a **proof of concept** (see **Sprint Informatics** on page 105).

...shrink your idea to its **minimum viable product** (see **Minimum Viable Product: Using the Startup Mentality** on page 149).

Sometimes, all it takes is time. You may need to wait for an opponent to retire, or public opinion to change, and then try again (see **Subterranean Informatics** on page 151).

> **The Lesson Learned**
>
> Even good ideas often fail on the first try. When facing defeat, your persistence is what will underlie your success.

Machiavelli on The Grand Gesture

> *...it should be considered that nothing is more difficult to handle, more doubtful of success, nor more dangerous to manage, than to put oneself at the head of introducing new orders.*
>
> —*The Prince*, Chapter VI

Informaticists are often considered "essential infrastructure," like water and electricity. And just like domestic utilities, we only get noticed when something goes wrong. But if people only think about you when they're angry and irritated, they are unlikely to become your supporters.

To reverse this, we publicize our victories—even our small ones. When we make a positive change for our docs, we celebrate it every time. But sometimes it's not about the small changes; it's about something big and exciting, something that will capture interest and enthusiasm.

Our Electronic Medication Optimization committee (which we joyfully call "EMO") works to remove low-value medication warnings that interrupt prescribing physicians. One frequently overridden alert is the "duplicate med warning."

At first glance, the warning seems like a good idea. It reminds the prescriber that a patient is, for example, already taking atorvastatin to lower cholesterol if they are about to be prescribed another statin from the same drug class. On the other hand, prescribers often combine doses of the same med to get to the right dose. When prescribing Adderall for attention deficit disorder, it is common to tell the patient, "take the Adderall five-milligram tablet together with a 20-milligram tablet." In such cases, the duplication is intentional, and the alert: "Warning, duplicate med!" is a nuisance. More than 90% of the time, the alert is overridden.

Suppressing this and similar alerts seems like an obvious move. Every mouse-click we eliminate helps our clinicians. And every time we find and remove such a nuisance, we broadcast and celebrate it in our monthly newsletter. We take credit for every improvement.

So often, informaticists work in the shadows. We listen to feedback from clinicians, eliminate small nuisances, and make doing the right thing easy (fewer alerts, more precise advisory language). Often, that "invisibility" is just fine. The planet continues to revolve, and we quietly do our work, making systems better, faster, and less obtrusive in the service of excellent patient care.

There are times, however, when a small change is not enough. For big problems and big frustrations, a big change that can be felt by many colleagues is needed.

For example, there is a rising tide of doctor and nurse burnout from administrative burdens (in large part due to EHR documentation and computer use). Often, this feeling bubbles up as frustration. Our informatics goal is to make a big improvement so that people can feel it in their daily work.

We were asked to address this burnout, beginning by reducing the task loads of nurses and doctors responding to clinical messages in the EHR. The "hated In Basket" is a frequent topic of grumbling complaints.

Our approach? Make a sweeping change. Why shoot for a 5% improvement just because it's "realistic"? Why *not* shoot for the moon? I refer to such a far-flung goal as a "grand gesture." Even if we fall short of our vision, the vision itself is galvanizing. Who *wouldn't* want to be part of eliminating 90% of the burden of the In Basket? Who wouldn't want to be on a project to completely eliminate passwords for prescribing medications in the EHR? Nobody, that's who.

We now have a culture of asking, what will get the physicians' or APPs' attention? Few people will notice when things get 5% better or faster, but big changes are perceptible even to busy clinicians. Here are some examples we implemented recently:

- We eliminated passwords for prescribing non-controlled substances (the result: 7 million fewer passwords typed per year).
- We removed automatic-forwarding of specialist charts to address a surfeit of low-value messages like, "I wanted to tell you that I provided routine eye care for your patient." "I wanted to let you know I saw your patient in the cardiology clinic and we made no changes."

Every week, a primary care physician received several hundred of these automatically forwarded specialist progress notes in their In Basket, even though only 3% of the messages had an important new diagnosis or change in plan.

- We committed a one-time purge of all In Basket messages *more than six months old*. We deleted 12 million messages. Individual physicians saw their "stored message" tallies drop by thousands as out-of-date and not-yet-deleted messages were purged.
- We began auto-deleting newly arrived In Basket messages after 90 days. We had discovered that almost all message threads were completed within three weeks, or their questions/topics were resolved in other message threads. Thus, messages older than 90 days were all clutter. This policy deletes over 100,000 messages a month.

Yes, incremental change is good, and you should celebrate it publicly so that your colleagues are aware that you continually advocate for them. But a grand gesture is seen and felt. A successful informaticist pursues both large- and small-scale changes. Without grand gestures, you are at risk of appearing irrelevant. Look for opportunities, and you will find them.

The Lesson Learned

Make small improvements and celebrate them publicly. But when you can, aim high and make "grand gesture" improvements that everyone can feel.

Machiavelli on The Burning Platform

> *...For injuries must be done all together, so that, being tasted less, they offend less; and benefits should be done little by little so that they may be tasted better.*
>
> —*The Prince*, Chapter VIII

Not everything you do will be celebrated. When you *must* change something that will be perceived as "bad," don't spread it out. Do it all at once.

Sometimes that "bad" thing involves disrupting a system that is disliked but tolerated. Radiology orders, for example, are complicated. To order a CT scan, you have to specify whether the test should be done with intravenous contrast, how urgently to schedule, what symptoms the patient has, why you're ordering the test, and what ICD10 diagnosis to use for billing. To save time, many physicians have created their own frequently ordered "favorites."

The good news: The EHR has exciting new capabilities. The bad news: To take advantage of these new tools, we have to wipe out every clinician's "favorite orders." This would force clinicians to rebuild their favorites from scratch. In other words, it will be extremely unpopular. The question: Should we wipe out *all* the favorites at once, or in smaller increments? Maybe we could wipe out CT scan favorites the first month, the MRI favorites next month, and ultrasound favorites the month after, so there is less work each time?

Our approach, informed by Machiavelli: *don't* get rid of things piecemeal—your users will hate it every time you make a change. Instead, rip off the bandage all at once. One painful change fades faster in memory than several smaller changes.

This concept has been referred to as a "burning platform"—a situation in which change is inevitable, and everyone must jump.

All-at-once was a successful tactic, and one we employed when circumstances forced our hand. A mandatory software upgrade, a change in best practice for a prescription, a change in federal regulation. Each of these forces outside our control was a burning platform. However, I never thought we could manufacture *our own* burning platform.

Someone who did understand the value of a burning platform was my mentor Dr. John Sbarbaro.

It was 1999. University Hospital was using 3M Clinical Workstation (CW) as a data repository. Most doctors used it to view their test results. Unlike the previous Stat-LAN lab computer, which deleted test results after 90 days to save disk space, the CW database was advertised as a "lifetime data repository" where results were kept forever. Groundbreaking!

At the time, about one in ten doctors dictated their daily progress notes, and the technology was clunky. Doctors would speak their notes into a handheld Dictaphone recorder, which was about the size of a stick of butter. The tape cassette could hold about 30 minutes of dictation, so you had to talk fast. Once full, the doctor would put the cassette tape in an interoffice envelope for the courier, who came every evening.

Imagine the logistics: You had to keep a supply of AA batteries for the handheld Dictaphones. Plus, every doctor who dictated needed cassettes and envelopes. The envelopes had to be labeled properly, and they all had to be collected by the daily courier.

Transcriptions would take up to two weeks, and then the courier would return with a sheaf of printouts. For hospital notes, these printouts had a peel-off sticky backing so they could be literally pasted into the chart. In clinic, the printouts would be punched with two holes to accommodate the metal binders of the paper chart. I remember seeing an industrial-strength hole punch that could handle three inches of paper at once. *That* was the pinnacle of medical records technology in 1999, when Dr. Sbarbaro taught me the power of the burning platform.

Given the limitations, it was not surprising that most doctors chose not to dictate. They instead scribbled on the paper progress note, peeled off the main medical records sheet, and kept a yellow non-carbon-paper copy for the clinic. Under that system, you knew your note was finished. You didn't have to worry whether your dictation would come back from the transcription service two weeks later. Hence the 10% dictation rate.

In the second half of 1999, two major changes occurred. First, the dictation company switched from cassette recorders to telephone-based dictation. Second, the company began delivering notes electronically alongside the physical printouts.

After some discussion, we realized these notes could appear in our data repository! But although this was a neat idea, I did not believe that putting notes in CW would help us much. After all, with only 10% of doctors dictating, most notes were *not* in CW, and you had to request the paper chart to see the complete record anyway. Until all doctors started dictating, electronic charts would be incomplete and unreliable. It seemed like an impasse.

When I told Dr. Sbarbaro about dictated notes and the electronic interface to 3M Clinical Workstation, his eyes got big. Like me, he realized that if *all* doctors dictated their notes, CW would become hugely valuable for patient care, storing all test results and all doctors' notes, replacing the need for paper chart delivery. The frustration of delayed and missing charts would evaporate.

Personally, I was pessimistic. I had learned that changing the behavior of busy doctors was really tough. Scribbling in the paper chart was fast—and 'fast' was favored. For most doctors, the paper system wasn't great, but it worked fine. They would not be enthusiastic about dictation.

Dr. Sbarbaro saw it differently. He saw only the potential benefits, and he was unafraid to create a burning platform to reach them. He announced to the doctors: "As of February 1, 2001, your progress note must be visible in 3M Clinical Workstation. If UPI coders don't see your clinic note in CW, we will not bill the insurance company for your visit." No note in CW? Sorry, you'd be working for free. This was obviously an unpopular piece of news. Dr. Sbarbaro also offered an olive branch: The dictation service was free to doctors. A benefit of over $1,000 a year, per doctor. Skip the typing, you can dictate!

I thought, "We can't do that, can we?!" Doctors don't change their behavior so fast; you have to convince them, and adoption of new tools is gradual. Such a bold move made me very nervous.

After the announcement, as I expected, there was a huge uproar from our doctors.

"This is unfair!"

"This is so much more work!"

"There is nothing wrong with the way we write our progress notes right now!"

Not to mention, "Pushing 30 buttons on the telephone is *not* why I became a doctor."

True, the dictation service at the time required a lot of button-pressing. You would first press "9" to dial for an outside line. Then dial the 10 digits for the transcription service phone number. Enter 6 digits for the date of service. Then the 5 digits of your physician ID code. The next 8 digits were the patient's contact serial number. Only then, the electronic document could reach the right patient's chart. For some doctors, these 30 digits were deemed an unreasonable effort to expend on every patient.

Dr. Sbarbaro held firm.

What would happen? Would doctors refuse? Would there be a revolt? A great uprising?

I held my breath that first week of February. There was a lot of grumbling. Then, there was a sudden interest in learning how to use the telephone dictation system.

When the numbers came in, physician progress notes in CW had gone from 10% to 99% overnight. Everyone, it turns out, wants to get paid for the hard work they do. And once cassette tapes and printouts were eliminated, turnaround time on dictation fell dramatically.

Within a month, it was clear that CW had become an indispensable tool. Dictation complaints faded, replaced by the rising demand for more computers in clinics. It was our first glimpse of what electronic workflow, central storage, and instant access could do in the years to come.

In hindsight, Dr. Sbarbaro's dictation mandate would have failed a year earlier, when the dictation interface didn't yet exist. It would have failed in the hospital setting, where it was less useful (the paper chart stayed at the patient's bedside for their entire hospital stay; missing notes and results were not a problem). The best burning platforms create a win-win where both parties come out ahead. In this case, the billing department got legible, timely notes, and physicians got a computer with immediate access to useful patient data.

Recalling this event years later, I manufactured my own burning platform. As described in an earlier chapter (see **APSO Notes, Maxwell, and the Meeting Before the Meeting** on page 98), I was working to persuade doctors to use the APSO note format. Despite my efforts, the new format had achieved only partial adoption. And then, an opportunity appeared: In 2011, we would be adopting the Epic EHR platform. This created a natural transition point. If I was going to shake things up by mandating APSO notes, why not do it from ground-zero, when installing a new EHR?

The adoption of a new EHR causes significant disruption. Everyone has to learn the new tools and develop new workflows. Within that large shift, I would include a smaller adjustment: I would invert the standard template for all progress notes from SOAP to APSO.

I told the white lie: "Sorry, our new EHR only comes with APSO notes."

With leadership support and a strong communication plan, Epic went live on September 1, 2011, and APSO notes became the new standard at our organization.

Did we succeed?

After the EHR switch, 100% of primary care physicians/APPs and 88% of specialists adopted the APSO format. This has been our documentation standard since then, and a growing number of hospitals nationwide have adopted the APSO format as well.[1]

> **The Lesson Learned**
>
> Chosen carefully, a burning platform can facilitate large-scale shifts in behavior. If there isn't a natural crisis, you can create your own.

What Is Your Vision Statement?

"Would you tell me, please, which way I ought to go from here?"
"That depends a good deal on where you want to get to,"
said the Cat.

—Lewis Caroll, *Alice's Adventures in Wonderland*

In 1997, I was given the newly created role of "Physician Liaison to the IT Department." A better name for it was "chief complainer." I earned my position by writing a seven-page screed about the terrible computer tools at the University of Colorado Hospital (see **Stat-LAN and the Chief Complainer** on page 42).

To fulfill this role, I would be excused from clinic duty one half-day each week to attend meetings about computer systems for clinical care. My main job description was "Explain all the great things that IT is already doing for the doctors." My job description did not include "represent physician viewpoints about the usefulness of IT tools in a busy clinical setting."

Over the decades, my non-clinician IT colleagues showed a recurring lack of appreciation for how chaotic clinical work is. Tools used in clinical areas must be streamlined and fault-tolerant. Clinicians are working quickly and multitasking to solve urgent clinical problems. Thus, IT or EHR tools need to be as simple as possible. Complicated tools that require extensive training fail in this environment.

On the other hand, my IT analysts, trainers, and colleagues often felt unseen: Clinicians rarely understood that technology is difficult to design well, and the human–computer interface was still a new idea, and not well tested. Our clinicians and our technologists would frequently clash over the simplest misunderstandings:

"We bought this big fancy EHR. So go turn it on and make the doctors and nurses use it." This administrator's view does not appreciate the pressures of clinical workflow and that every specialty has unique requirements.

"Just put another hard stop there. That will make the doctors and nurses do what we need." This is the view of a misguided leader, thinking that the EHR is a hammer to enforce a new policy or raise the percentage on a new quality metric.

Effectiveness is a balance between clinicians having uninterrupted, efficient flow, and clinicians being gently nudged by EHR suggestions. The best nudges are gentle and well-timed, making it easy to do the right thing. However, such gentle, effective nudges are hard to design.

Once we had our Sprint training model (see **Sprint Informatics** on page 105) up and running, we heard requests like this: "I see that the clinicians have learned a lot from the Sprint optimization team. We would like you to teach this new billing module during Sprint and get them to use it more effectively. It only takes a few more minutes per patient." This is a finance leader's view that does not appreciate the informatics focus on reducing the EHR burden.

"The Sprint team should just focus on the doctors; they're the ones most burned out. Skip the nurses and staff." This was my own misguided idea, not appreciating the value of different team members on providing efficient care.

So many well-meaning yet misguided ideas. What they lack is a set of guiding principles. Having guiding principles for a working group, like our informatics team, helps to ensure that everyone on the team is driving in the same direction and has the same goals. What then, would be *our* vision statement?

Over the years, the team has grown to 25 physician and provider informaticists. I reverently call our group "The Large PIG." PIG stands for "Physician/APP Informatics Group," and it's "large" because it's the entire team. Not everyone loves this name. Sorry, when you become the boss, you can change it.

Our informatics group serves a large health system with many competing departments wanting to influence physician/APP behavior. Increasingly, they have realized the EHR is the nexus through which flows all patient care directives. Thus, influencing clinician behavior means influencing the EHR design.

The compliance department wants to reduce risk of regulatory violations by including protective language and behavior reminders for physicians. The legal team wants written disclaimers to reduce lawsuit risk. The billing office would like all possible data inserted in the note to reduce loss of revenue. The quality mavens would like all important quality benchmarks mentioned in the progress note. Physicians would like just a *simple and brief explanation* of what happened at the previous visit and what the physician was thinking at that time. The Large PIG is at the wheel, but we have many backseat drivers. With all these conflicting directions, how can we avoid steering the EHR off the road and into a ditch? This is why you must establish a unified vision where all members drive in the same direction.

In meetings about our vision, we debated our primary goal: To help overburdened physicians and APPs by designing a better EHR. We knew we also had to satisfy our compliance, legal, billing and quality teams, but our first duty was to our physicians.

Our first draft was, "As physician informaticists, we seek to improve physician effectiveness by building world-class relationships, reducing the EHR burden, and improving patient satisfaction using innovative tools." With this as a basis, we argued extensively over the principles and the wording.

For example: What about teamwork? We should address physicians, APPs, nurses, and clinical staff. We made it "physician and team effectiveness."

But wait, informatics effectiveness comes from building great relationships. It was crucial to grow a multispecialty team representing all regions and most specialties. Also, we preferentially hired excellent communicators with strong conflict resolution skills over tech-savvy doctors. It is much easier to teach technology than to train communication, consensus-building, and leadership skills (see **Grow Your Team: Hire the Techie or the Leader?** on page 163). But we knew there had to be a reference to "tools." Too often, IT and informatics are thought of as "the mechanics who fix things," when instead we listen, diagnose, design, gain consensus, and *then* create. So, we kept "innovative tools" but moved it after the more important "relationships."

And oh, right—"wellness." We changed our wording several times. "Reduce EHR burden," then "increase joy," and then "increase efficiency," and then "reduce burnout." Ultimately, we chose the complementary balance of both "wellness and effectiveness."

We then swapped out "world-class" (a colleague astutely asked, "What does 'world-class' even *mean*, CT?") for "extraordinary." We decluttered our language. We added "APPs" to our name, because advanced practice providers are increasingly important in our organizations and in our leadership. Was that everything? We sat in contemplative silence. Evidently it was.

Here is our final Large PIG Vision Statement:

"We improve physician and team wellness and effectiveness by building extraordinary relationships and innovative tools."

The Lesson Learned

A cohesive vision statement will ensure your team always knows its priorities. And crafting a vision statement together is itself a valuable teambuilding activity. When there are difficult decisions and many competing demands, a strong vision statement can clarify team direction and serve as a powerful North Star.

Three-Circle Venn Diagram: A Personal Approach to Influence

"Don't ask what the world needs. Ask what makes you come alive, and go do it.
Because what the world needs is people who have come alive."
—Author and philosopher Howard Thurman

How do we choose what to work on?

EHR design presents vast possibilities. It affects both patients and clinicians, and there are hundreds of processes that could be made more efficient. There are always cool new tools to consider. The list of potential projects is overwhelming, and informaticists see many more opportunities than they can reasonably pursue.

One of my colleagues was debating how to choose their next project. Should they develop an inpatient EHR Sprint optimization team? Or take over the leadership of the Physician Builder program? They also had the option of building specialty care pathways and subspecialty flowsheets. Was it better to pick "low-hanging fruit" or reach for a big, important project with high visibility?

Google famously granted their employees "20% time." This meant that each employee could spend 20% of their week on a personal passion, using their knowledge and connections to tinker with a new idea, with the condition that they periodically present their work to their boss for consideration. This encouraged their smart employees to stay engaged in their primary work while exploring personal initiative. And this personal initiative often bore fruit! For example, Google's Gmail client was one such side project by a Google employee who wanted to build a better email system.

How might we apply this core idea? Unlike Google employees, no one in healthcare has "20% time" to spend how they please. But we do have to allocate our time somehow.

As with the Google employee who created Gmail, I have noticed that projects are generally more successful when they are a deep personal interest. Certainly, informaticists have the skill to work on almost any project in clinical medicine related to the EHR. However, when we can align a project with our personal interests, it's a win-win. The individual tends to think about the project in their off-hours, and they bring more joy, creativity, and energy to the work.

Since noticing this, I have done my best to allow my informatics colleagues to pursue their assigned projects *and* consider their personal interests. Where possible, I want them to tell me what their passions are. And then I want them to parlay that passion into a concrete project that will conjure something cool into existence.

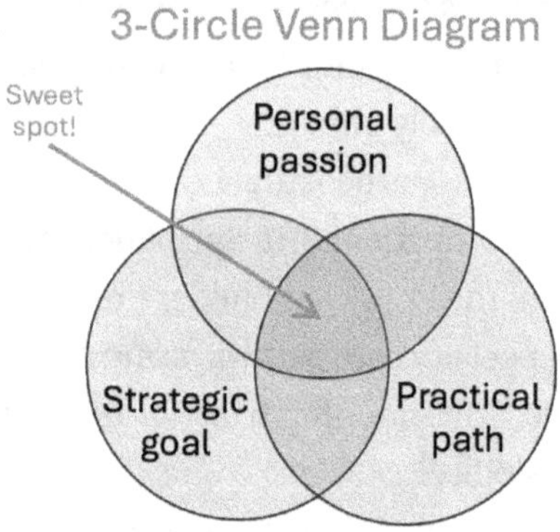

Since implementing this rationale, I have noticed a second pattern. The best passion projects land in the center of a three-circle Venn diagram that includes passion, a practical path and a strategic goal. These projects are most likely to succeed.

Let's deconstruct this idea.

Passion Projects: This is any project that makes you think "I would love to work on that," or "Wouldn't that be cool?" For example, maybe you want to build an Annual Wellness Visit note that simplifies the visit with patients and *makes the right thing easy* by generating all the relevant CPT2 codes. Or maybe you want to teach your colleagues to use Anticipatory Guidance to reduce patient anxiety when test results release to them immediately. Fill in your own idea here. The important thing is to ask yourself, "What do I care about?"

Strategic Goal: From the above examples, the Annual Wellness Visit improvement might be part of your organization's push toward performing well in a value-based care program. Great! It should be easy to find leaders who are friendly to your idea and will give you support and funding. On the other hand, maybe Anticipatory Guidance is not a system priority. This makes your project harder to achieve. When choosing a heading on which to launch, you're more likely to reach your destination if the wind is already blowing in that direction.

Practical Path: Some ideas may involve working with a vendor or someone in another department who doesn't share your priorities. In other cases, maybe you need only yourself or a good friend who will work with you, and no one can veto or get in the way. For example, suppose the CPT2 code automation required analysts from the finance department who won't get to it until next year. That would be a bad project to start work on. However, if you let go your dream of automated billing codes, the Annual Wellness framework would still be a big win for primary care colleagues, and you have the physician builder skills to complete the work yourself. By reducing the scope of your idea, you can keep your passion project's strategically aligned goals and create a practical path forward. That's a good project!

Projects that live at the intersection of all three circles are the ideal. They are likely to succeed. Projects in zero or only one circle should not go forward. Projects with two circles might have a chance, particularly if you have an overriding passion and some strong allies.

The Lesson Learned

We are all busy, and we all have countless goals to chase. Find a passion project, one that is aligned with system strategy and has a practical path forward.

The Ferret and the Flywheel

Ferret Frenzy is a big problem for organizations, particularly in informatics.

Unlike the history of medicine, which goes back millennia, and the history of medical specialization, which goes back centuries, informatics is based on computing power and the flow of information. As an art and science, it is only decades old. Everything we do is new and constantly changing.

Even worse, Moore's Law stipulates that computing power doubles every 18 months while its cost drops by half. We have lived in Moore's world for over 60 years. The smartphone in our pocket has more computing power than did the entire NASA Space Agency at the time of the Moon landing. It seems that every few weeks there is a new breakthrough technology, a shiny new object. Like an easily distracted ferret, our heads turn to follow each new technology we hear about. I call this "ferret frenzy."

Examples of Ferret Frenzy:

1. We hired an army of scribes to type in the EHR and improve efficiency. This effort lasted a few years and has now faded.
2. We insisted on 100% computerized physician order entry (CPOE) for patient safety. This lofty metric has since been redefined and largely set aside.
3. We implemented speech recognition for everyone. It was adopted by two thirds of our clinicians. This Ferret-driven idea ended up being a big success over the last 15 years. Sometimes we are lucky and a shiny object becomes an important technology for your organization.

In the book *Great by Choice*, Jim Collins and Morton T. Hansen talk about the 20-mile march and the discipline of the early Antarctic explorers.[1] Some explorers would hike across the tundra on good days, crossing many miles, and then hunker inside the tent on bad weather days, sometimes stuck for a week at a time. In contrast, the most successful explorers set themselves a goal of marching 20 miles every day, regardless of the weather. This 20-mile march showed two kinds of discipline: not overtiring themselves on the good days, and also getting out of the tent on the bad days.

In a similar vein, Collins wrote about the "virtuous flywheel." Pushing steadily and gradually in the same direction, an organization makes boring but consistent progress. By maintaining direction, the flywheel gains momentum, and future changes in the same direction get easier. This momentum leads to high performance.

In both cases, Collins describes consistent progress as the root of successful companies. But informatics is a constantly changing field with many new tools and technologies to consider. How do we reconcile the ferret and the flywheel?

First, let's recall the "Diffusion of Innovation" framework by Everett Rogers.[2] This describes the technology adoption curve for organizations and divides that curve into innovators, early adopters, the early majority, the late majority, and laggards.

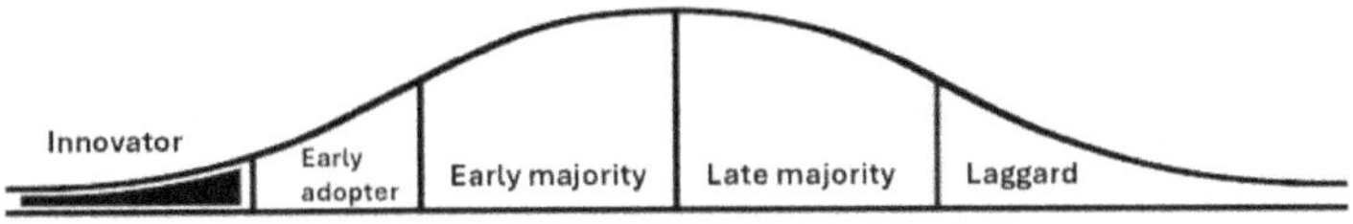

People who are quick to pursue new ideas and directions (let's call them "ferrets") are perceived very differently along the technology adoption curve. Let's see how ferrets fare in different situations.

Innovator organizations may have a ferret in a leadership role. As leaders, they place big bets on companies with exciting new ideas. They expect frequent failure and occasional big successes. They're the most daring and adventurous ferrets.

Early Adopter organizations encourage their ferrets to scan the news for innovator successes. They are willing to pay a premium for early access to immature technology. They are willing to be the "first follower" and accept failures in the hope of getting in on the ground floor.

Early Majority organizations are spurred by their ferrets to pay higher prices on newer commercial products that have been through the initial innovation cycle. These organizations are a little cautious, but not conservative in their decisions.

Late Majority organizations have disappointed ferrets. They purchase new technology once it becomes an established, low-cost commercial service.

Laggard organizations rarely hire ferrets. The ferrets who do work there gaze sadly at the landscape of new technologies and resign themselves to reading about other ferrets doing cool things.

Of course, none of these approaches is "correct." Each has associated risks and benefits. But when it comes to adoption readiness, I propose that an organization's decisions have three underlying components: their leaders' attitudes toward change, the strength of the informatics teams, and their budget for innovation.

Org type	Innovator	Early Adopter	Early Majority	Late Majority	Laggard
Change-ready leadership	Strong	2 of 3 must be Strong	1 of 3 must be Strong	Chaotic mix including Medium	Not ready
Physician/APP informatics	Strong				Not ready
Budget for innovation	Strong				Not ready

Innovator organizations have strong leaders, capable informatics teams, and big budgets. Laggards are the opposite: their leaders, informaticists, and budget are insufficient to manage change. The other categories contain a spectrum of maturity in these areas.

Keep in mind that even within a single organization, the adoption curve can differ depending on the specific topic at hand. For example, an organization that is a financial technology innovator may also be a clinical technology laggard. And in hindsight, I have learned important lessons from working in each category of adoption readiness.

For example, in the early 2000s we were part of the Late Majority in acquiring an enterprise EHR. We were using many departmental EHRs: Pulsecheck for emergency care, CRIS for psychiatry, Xcelera for cardiology, McKesson for the hospital, Allscripts for clinics, and 3M Clinical Workstation to tie it all together. We dreamed of investing in a big enterprise EHR (Oacis, Eclipsis, Cerner, or Epic), but we didn't have the budget or the leadership interest. We watched as other organizations purchased and deployed enterprise-sized information systems and benefited from reduced handoff errors and seamless information transfer. And we were envious.

As an individual, your own interest in innovation cannot override what your organization is ready for. In this Late Majority example, my organization was in no position to adopt an enterprise EHR. Knowing what I know now, the lack of budget and leadership interest should have tipped me off. I could have saved myself the frustration of pushing for an EHR that could not exist in my organization at the time.

Twenty years later, on the other hand, our organization was an Innovator. We had deployed an enterprise EHR (Epic) that brought together all specialties, hospitals, and clinics. Our leaders were innovation-focused, the budget for innovation was substantial, and our informatics team of two dozen worked closely with operational leaders.

When the COVID pandemic lockdown began, we needed an innovative way to train hundreds of people quickly and remotely. We had already invested in two innovation companies: uPerform and Amplifire. With these tech partners, we rapidly built a successful asynchronous, self-paced learning plan and trained hundreds of outpatient nurses and doctors to care for hospital patients in record time. This system has since been adopted by many other organizations.

As with so many things, balance is the key. An organization of disciplined leaders will use consistent, incremental changes to improve quality and reduce clinician frustration. Small improvements spin the virtuous flywheel. But when a new technology or new opportunity comes along, you must also know when—and when not—to pursue it. Strategically minded leaders avoid the ferret frenzy.

The Lesson Learned

Innovation is hard. The rewards can be great, but so are the risks. Know where your organization stands on the tech adoption curve. This will tell you when to focus on turning the flywheel and when to let your inner ferret out.

Minimum Viable Product: Using the Startup Mentality

Your project keeps getting stuck in committee. Or maybe it can't get funded, or needs too many permissions, or just isn't finding its footing. You're frustrated. How are you supposed to get anything done?

An Annual Wellness Visit (AWV) is an important meeting between a primary care physician and their patient for two reasons: First, it improves care by prioritizing prevention and wellness, goals that are often missed when patients come in for an urgent problem or to manage a specific disease like diabetes. Second, insurers measure a primary care physician's "quality of care" by the percentage of patients over age 65 who have completed an AWV.

There are dozens of tasks that physicians must perform and document successfully in order to bill for this code in an outpatient primary care clinic. The intent of the AWV is to set aside one visit per year to assess the patient and plan for the coming year, setting long-term goals and closing care gaps.

Our big idea: if the EHR did a good job of spotting the care gaps and made it quick to order the easy things, then the physician could focus on important conversations, like discussing advance directives or helping patients build healthy habits.

The problem was, we couldn't agree on what to put in our AWV tool.

Despite the pressing need for an EHR template, we would meet monthly with primary care docs and arm-wrestle over what *ought* to be in an Annual Wellness Visit (beyond the bare minimum mandated by Medicare). Even the bare minimum AWV has 17 items, including screening for cancers and heart disease, discussing vaccines, discussing end-of-life planning, and encouraging disease prevention.

On top of this, our conscientious colleagues wanted to include metrics like "patient stated goals" and "exercise minutes per week," and more besides. Soon, we had doubled the requirements for the AWV template. Too much had gone in, but what could come out? We debated for months with no end in sight.

In situations like this, it is worth stepping back to ask: What is the smallest possible version of your idea that will work? If you're getting bogged down in details, keep your project small and complete it. Deploy it and get it into general use. You will learn a lot and grow from there.

My colleague, the physician builder in charge of the meetings, was trying to wrestle the group to a consensus when I suggested: "Just build the minimum viable product. We will push the MVP through committee, with the promise that it meets regulations. *Then* we can go back and improve it by adding some highly desired items."

This was the kick that finally moved the ball over the goal line. Once we trimmed off all the "fat" from our discussion, we could agree that the bare bones AWV met all the basic criteria, and we could all start conducting these visits with an efficient template, making documentation easier for medical assistants *and* doctors, thereby reducing the burden of completing these important visits and, in turn, improving patient care.

The Lesson Learned

Startup companies laser focus on identifying their MVP. Too few features in a new product, and it is not useful. Too many features, and the cost of development slows things down. MVP is the sweet spot. In other words: Keep It Simple, Stupid (KISS).

Subterranean Informatics

Are there times when you feel like your organization doesn't support you? Do you regularly encounter insurmountable obstacles? That no one understands what informatics can do? *Are your colleagues all idiots*?

If these questions resonate, then perhaps subterranean informatics is for you. Sometimes, you are a team of one. Sometimes you need to find what small actions you *can* take. Look inside yourself and challenge your own assumptions.

Here are some obstacles I was unable to overcome:

In 2000, I was promoting patient access to test results and progress notes and getting nowhere. Clinical leaders were not supportive (see **The Naïve Scientist Part 2: "But, the P-Value!"** on page 193). My project came to a stop. I took this as a personal failure.

In 2004, I was pushing for computerized physician order entry to improve patient safety (see **CPOE and the Critical Bug** on page 159). The technical barrier was too high. The project stopped. Another failure.

In 2013, we attempted to simplify "radiology indications of use" for radiology orders (see **Fall Down, Get Muddy, Do Informatics** on page 235). Built with the available technology, physicians found the long-scrolling lists unusable. A week later we removed the RIU tool. Failure.

Publicly, these projects came to a halt. Privately, I was still looking for a way forward, a clandestine way to continue informatics work to serve my busy clinical colleagues. If you can't climb over the obstacle, dig a tunnel and go underground. Be out of the spotlight and do your own thing that does not require official leadership support.

Here are four subterranean paths: learn, pivot, allies, and time.

Learn. Turn on your learning mode. Break the box you're in. Read like hell. Ask smart colleagues for advice. Practice new leadership skills.

Pivot. No one says you have to stay in your lane. Go explore! You know your company's culture. There are always opportunities.

Allies. Look for members of your organization who align with your goals. Together, you have a better chance.

Time. As time passes, naysayers retire. Technology improves. Culture shifts. Bide your time and try again.

Using these strategies, I dug tunnels under my original obstacles.

In 2000: Patient access to test results failed. I **learned**. I read books and learned from colleagues about change management. I tried again and succeeded in 2016.

In 2004: CPOE failed. I **pivoted**. An immediate opportunity arose to help nurses improve safety with bar code med administration. I attached myself to that project and helped it succeed.

In 2013: RIU failed. I found **allies**. The finance and radiology departments had separate but aligned reasons to deploy RIU. We redeployed years later, with better technology and savvy colleagues.

In the intervening years, while waiting for the chance to restart big projects, my subterranean tunnels helped me find success in smaller projects: improving order preference lists, publishing our experience with speech recognition in the exam room, pioneering the sharing of radiology images with patients.

The Lesson Learned

When you feel like you have failed, you can go subterranean. Choose a tunnel to go under your obstacles. Learn, pivot, find allies, let time pass. Re-emerge and carry on.

Don't Be Selfish, Be Useful

I joined the faculty of the University of Colorado School of Medicine in 1995, starting as an assistant professor. I was hired by the recently promoted Division Chief Bob Anderson. Observing his leadership style, I realized he was a data-driven, no-nonsense kind of guy.

In meetings, his tendency was to lay out an agenda, give a brief explanation of the topic, take questions, and then move on. If any member raised concern about an issue in clinic, or some other deficiency, his response was to immediately jump to problem-solving, often involving the person who had brought up the concern.

Among my colleagues at the time, this was a new experience. Our previous leader would often allow concerns to be raised and briefly discussed, but with no expectation of immediate action or resolution. With Bob Anderson, it was different. We learned this lesson when a faculty member voiced their thoughts about the lack of reflex hammers in clinic: "Bob, I find that not every room is stocked with a reflex hammer for neuro exams."

"John, good point. What is your solution?"

"I just wanted to bring this to your attention."

"Very good. Why don't you check with the clinic manager and see what it would take to get the hammers purchased and stocked and let us know next meeting what you find out. Thanks."

"...Okay."

Quickly, it became apparent that "Bullet Bob" did not permit empty complaints. Don't say anything unless you were willing to back it up with action. But I must have been slow to digest this, because I had the brilliant idea of raising my hand:

"Dr. Anderson, looking at the patient education materials in clinic, it looks like some of them have been photocopied many times and are nearly illegible."

"CT, good point. What is your solution?"

"I… could go through some of the files and make clean copies of the more frequently used ones."

"Good idea. Our faculty practices at five general internal medicine clinics in Denver. There's one at University, two at Denver General, one at Presbyterian, and one at the VA. Why don't you index patient education materials at all five sites, find the best ones, create a master notebook, and then distribute them back to all five clinics? Come up with a budget and timeline on how quickly you can complete the project, and I will approve it. Next item of business."

Just that quickly, I had a new task. It seemed like a lot of work on top of a full clinic schedule and hospital rounding duties as a general internist. Not at all what I raised my hand for. On the other hand, being a junior member of the faculty, I was keenly aware that I did not yet have a reputation. This was an opportunity.

I spoke with Dr. Anderson after the meeting. We agreed he would cover the cost of office supplies, including three-ring binders and sheet protectors, and the cost of photocopying. He asked me to complete the task in two weeks. I agreed.

Although I had initially groaned at receiving this menial task, I actually enjoyed driving around to all five clinics, meeting colleagues across town whom I had never met, reviewing reams of patient education handouts, many outdated and some nearly illegible, in order to compile a master notebook of the best education handouts. At the end of two weeks, I had created a master file of about 200 patient education topics, retyping many of them into new, legible handouts, which I gathered into five identical three-ring binders, indexed and organized in sheet protectors and dividers, and delivered them to each clinic. The final copy, I brought back to my own clinic at the university and placed it on Dr. Anderson's desk. Done! On time, and within budget.

As a result of finishing this simple project, Dr. Anderson saw me in a favorable light, and in the coming years, he assigned me to important and rewarding tasks, including my assignment to a "computers in medicine committee" a few years later. I learned to be useful as an underling.

I recognized how my boss thought and was willing to take the extra step when asked.

As a leader, there is an important flip-side lesson: I have learned to reward colleagues for their initiative when they see a problem and tackle it on their own.

In a talk called "We Will Win and You Will Lose," Panasonic founder and business giant Konosuke Matsushita famously remarked that US businesses relied only on managers for good management ideas, whereas at his company, they encouraged the ideas of all team members. He stated, "The survival of firms is so hazardous in an environment increasingly unpredictable, competitive, and fraught with danger, that their continued existence depends on the day-to-day mobilization of every ounce of intelligence." And thus: "We are going to win, and the Industrial West is going to lose out."[1]

Similarly, I often hear great ideas from my direct reports. I encourage them to find stupid stuff and do something about it. I give them full credit when they succeed. It's a win-win tactic; encourage initiative from everyone and avoid "waiting on the boss."

The Lesson Learned

As an underling—first, be useful, good things will come. As a leader—reward your direct reports for initiative and good work. You need good ideas from all the minds.

The 70% Rule for Decision-Making

Two decades ago, I experienced a crisis at my organization. Our new CFO (chief financial officer) was causing a stir. He made an impact right away, tightening our operational and financial belts.

This CFO saw that we were losing money each year, and he immediately began looking for ways to address it. One opportunity he saw: creating consistent criteria for how each clinic was staffed. There was no standard for how many nurses and medical assistants were needed per patient. What he implemented was a "volume-based staffing model" using national benchmarks. "If you see *X* number of patients, then your clinic may have *Y* nurses and *Z* medical assistants." These changes took effect immediately. Some clinics lost medical assistants and telephone staff; in a very few cases, a clinic added a staff member. Many people were let go.

The organization was in an uproar. Patient satisfaction plummeted as our telephone "speed to answer" lengthened by minutes and our clinics scrambled to reorganize how to room patients and coordinate their care. As a result, several medical directors resigned, prompting additional chatter and among colleagues. "It's not worth working here anymore. It's clear that our leadership doesn't care about us."

I pondered my own resignation, too. Early in this CFO transition, I had many questions but few answers:

- At that time, EHR use was optional. Would our EHR project be canceled?
- What does this CFO think about the EHR and about me in particular?
- Was my job on the chopping block? Should I just leave before I'm fired?

We both know now that I stayed at my organization, but you can imagine how uncertain my position was at that time.

If you've lived through a leadership change, perhaps this sounds familiar. Perhaps you've experienced how the actions of a new leader significantly impact daily operations. In some cases, the transition is seamless—the new leader fills the shoes of their predecessor and continues policies that are working well. In other cases, a new arrival shapes the organization into a new vision. Depending on the individual's skill, these changes can cause major disruptions.

How, then, did I make this decision? In my case, I was an inexperienced leader. I knew that I could always return to seeing patients as a physician and give up my protected time as a clinical informaticist. Informatics was so new at the time that the nearby hospitals didn't have designated physicians leading the work on electronic health records. Despite the uncertainties, I stayed put. Truthfully, I had no framework to think about this decision, and it was just luck that it was the right one.

Only you can decide about the universe of information you need to make a decision. And waiting for "100% of the information" means that you know how everyone is thinking about this situation, including the CFO, the CEO, your medical colleagues, and your nurses. It means that you know what decisions are about to be made to adjust clinic operations. It means you know if your job is secure and whether you will have more or fewer resources to work with in the coming year.

There is one problem: You never know 100% of this information.

Waiting until you know 100% means you have delayed too long and lost an opportunity. It means someone else may have scooped you and taken over, or you may have lost an opportunity to negotiate a better situation.

In the years since, I've learned more about decision-making. Colin Powell famously wrote in *My American Journey* that, when dealing with uncertainty, it is important for a leader to make a decision once they have between 40% and 70% of needed information. I find these percentages to be a helpful guide.

For the purpose of decision-making, 40% information is the bare minimum. Having less than that means you may jump prematurely. Perhaps you decide to leave before you're fired. Later, you learn that every leader who stayed received a major retention bonus and that the new CFO ramped up investment in electronic health records. Furthermore, the health system decided to sweep away the sandbox collection of many small departmental systems in favor of a fully integrated solution in Epic, that fabled system that everyone wants but can't afford. Indeed, this is what actually transpired the following year. If I had chosen to jump ship, I would have sorely regretted it.

In my two decades of informatics leadership, I've worked under four CEOs. Each new CEO reshaped the organization. Changes were inevitable. I have also mentored a dozen informaticists in other organizations about similar career decisions. In some cases, the right decision was to leave, in others, it was to stay. In some cases, both options were reasonable.

As an internal medicine physician treating medically complicated patients, I consider myself a scientist. I base decisions on clinical history, examination findings, laboratory test results, and my deep knowledge of pathophysiology. I tend to wait until 80% or 90% of the data is known before starting a treatment or prescribing a medication. Any uncertainty puts patients at risk.

On the other hand, in my informatics leadership role, there are no "blood tests" or objective exam findings to bolster my decisions. I have to make decisions based on mixed opinions and incomplete data. It is uncomfortable to decide quickly. Yet, the past decades have taught me that waiting too long to make a decision is itself a bad decision. Waiting has consequences.

> **The Lesson Learned**
>
> In decision-making, timing is difficult. Take guidance from smart leaders outside of healthcare, and wait only until you have *enough* information—not all of it. The 70% rule reminds us not to wait too long.

CPOE and the Critical Bug

In 2004, it was still the early days of computer systems for hospitals and clinics. The computers were much less powerful, and our knowledge of user experience and the human–computer interface was still rudimentary. Rather than choosing between different EHRs or fine-tuning specific tools as we do now, informaticists were often spearheading the leap from analog to digital—paradigm changes with broad implications. With limited funding, that sometimes meant making difficult decisions.

One such leap from analog to digital was computerized physician order entry (CPOE), which would allow busy doctors to use keyboards and mice to specify what tests and treatments patients should receive. The new tech promised to reduce errors related to illegible handwriting, and University Hospital wanted to be on the cutting edge.

Our organization was part of an early adopter group for McKesson Health IT, a six-organization collaborative. McKesson had purchased intellectual rights to WizOrder, a CPOE system developed by Vanderbilt University.

We were all excited. Vanderbilt's informatics team was famous, and their WizOrder product had an amazing search function. Standard CPOE software required lots of clicking and selecting to order a single medication. For example, just to order "metronidazole, 500 milligrams, intravenous, administer every eight hours, start immediately," you had to go through a lengthy process of clicks through many menus (prescription: antibiotic class: metronidazole: quantity, milligram dose: method of admin: frequency: start time). Let's hope you didn't click something wrong.

In contrast, WizOrder could discern your intent from typed abbreviations. If you wrote "metr 500 iv q8," it would return "Metronidazole, 500 milligrams, intravenous, administer every eight hours, start immediately."

Naturally, we *wanted* it. WizOrder would make the difference between a technology that was adopted begrudgingly and something doctors would want to use. The changeover would be simple and fast.

Except, nothing is ever simple and fast.

We purchased an early adopter license, and a team of analysts spent six months installing and testing the software. Then, as the go-live date drew near, the analysts discovered a vital software glitch.

All the major components of CPOE were functioning: ordering blood tests, radiology tests, medications taken by mouth, nursing interventions. Ninety percent of the tools transmitted seamlessly to the receiving departments. Ninety percent wasn't perfect, but we were ready to streamline hospital orders and say goodbye to illegible handwriting and internal faxing delays. The new CPOE systems could reduce time from ordering to administering treatments by up to 90 minutes. That speed was crucial. Speed saves lives in hospital care.

But then we learned about the software glitch in IV medications. Somehow, the system multiplied milligram doses times milliliters of IV fluid to create nonsense doses for our sickest patients. It was a significant safety issue. What to do?

Should we:

A) Push ahead and tell our pharmacists to "just be careful and correct all the IV orders"? (This would dramatically increase pharmacy work.)

B) Push ahead but keep IV medication orders on paper/fax process? This meant that the highest-risk IV orders would still have the same paper-based errors of illegibility and no dose checking.

C) Pause the project until the vendor fixes the glitch? (This would take an estimated 6 to 8 weeks, with the team left idle.)

D) Pivot away from CPOE entirely and install a different software module for a no-cost license swap?

Our other software option was Bar Code Medication Administration (BCMA), a patient safety initiative to reduce medication errors. Our nurse leaders really wanted it. It would ensure accuracy of the Five Rights of Medication Administration: right patient, right drug, right dose, right route, and right time. The barcode scanner was wielded by the nurse who would scan the med, scan the patient, the computer checking the exact time. All the safety checks are processed in milliseconds. If the med was too soon, or perhaps for the wrong patient – BUZZ. "Alert."

The debate was fierce, but ultimately we chose option D. We pivoted to BCMA.

As a result, our BCMA project became the first installation in the Rocky Mountain multistate region, and was a big win for our nurses and hospital. The safety checks reduced medication errors dramatically. It was a successful deployment in every sense.

We imagined our vendor would fix the IV dosage problem quickly, and we would get CPOE funded again the following year. However, instead of six weeks, it took over a year to get the bug corrected, and by then, our hospital's funding priorities had changed. The CPOE project stalled. The hospital physicians continued to scribble and fax paper orders with illegible writing for eight more years. Meanwhile, the hospital nurses were using computer charting and barcode scanning to improve patient safety.

In contrast, Duke University, another early adopter, went forward despite the bug and successfully deployed McKesson CPOE. Their organizational culture and pressures were very different. The year prior, a critically ill organ transplant patient had died because of a medication error that could have been caught by CPOE. They were laser-focused on eliminating future medication errors. Later, when we asked how they overcame the IV software glitch, they told us: "We hired 50 more pharmacists. Funding is not a problem in our pursuit of safety." These pharmacists were dedicated to receiving and painstakingly reconstructing every IV med order by hand.

The Lesson Learned

Leadership decisions are not made in a vacuum. Organizations have different finances, pressures, and internal cultures. Even the same technical glitch can lead to a different leader response, depending on the context.

Grow Your Team!

Becoming a leader means having direct responsibility for other people. Being a good leader means knowing how to guide and support your so-called "direct reports." An effective team is more than the sum of its parts, and a strong team will accomplish more and greater things. The bad news: Unless you develop your leadership skills, your team could descend into chaos and become ineffective.

Being a good team leader means choosing the right people for the team, assigning them work that matches their strengths, and helping them become more capable, more responsible, and more fulfilled.

Grow Your Team: Hire the Techie or the Leader?

As you grow your team, how do you choose your teammates? When I began hiring people, I learned an important leadership lesson from the School of Hard Knocks.

In academic medicine, people are promoted on the merits of their individual work—papers published, teaching evaluations, and patient satisfaction ratings. In non-academic organizations as well, physicians who are great at seeing patients are raised up as leaders. Promotions usually come with expectations: You're asked to chair a committee or lead a team. And yet, training in leadership skills is usually overlooked. The assumption is, "Since they're great with patients, surely they're a good leader."

This assumption is wrong. Here's how I know.

In 2004, we implemented Allscripts TouchWorks EHR in a rheumatology clinic. The clinic nominated a junior faculty member to be the specialty subject matter expert (SME). This recently hired person was still growing their practice; they were computer savvy but not yet clinically busy. Seemed like a good idea at the time.

Not knowing any better, I was pleased to have any volunteer from the clinic to help build content for their specialty. In a few weeks, this faculty member built dozens of note templates to address the diseases seen in rheumatology. I was delighted. After all, wouldn't rheumatologist-built templates be eagerly adopted by other rheumatologists?

Well, no. At go-live, the rheumatologists used only a couple of generic note templates; their 100 specialty-specific templates went ignored.

As a result, everyone felt we had wasted our time. The physician who had built the templates felt their effort was meaningless, the rheumatologists felt that the "junior guy" was too new and had not built consensus around the EHR tools, and our IT analysts had helped to support tools that were going unused.

There are several failures in this story, and the first is my own. I had chosen a specialist to be the subject matter expert, assuming they would also be an excellent leader, communicator, and consensus-builder. They were not.

I had failed as a leader, not appreciating that technical experts are not automatically respected by their colleagues. I could have observed closely, created a buddy system with a more senior leader, or coached them on the importance and tactics of consensus-building.

Also, I could have chosen better.

An EHR specialty champion must be clinically respected. They should *not* be "the new person who isn't busy," nor "the junior person who knows about computers." Clinical respect translates into trust among colleagues and a willingness to adopt the tools built by this trusted person.

Typically, this means someone with prior leadership experience. Ask interviewees to describe how they resolved a major conflict when leading a project. A surprising number of colleagues interested in EHR and informatics work have no such experience. Well-developed communication skills and experience navigating conflict and are crucial. If you have a choice of tech-savvy candidates, select the candidate with prior leadership experience.

Sometimes there *isn't* a strong candidate for this work. In those cases, there is no easy answer. You will have to choose: Select the tech-savvy candidate and train their leadership skills? Or select the respected clinician and give them technical training? The second choice is better. Where necessary, it is relatively easy to teach technology skills. It is much harder to retrain an uncommunicative genius.

Years later, a similar hiring situation arose, and I chose differently. Approaching a medical specialty practice, I asked for a subject matter expert from the specialty leaders. They put forward several candidates, one of whom had deep experience in technology but unproven communication skills and no leadership experience. The other was the medical director of a small clinic who had an interest in informatics but little tech experience. This time, the choice was straightforward: the clinician with leadership experience.

This person went on to technical training and has since become an invaluable agent of change, not only for their specialty but in many other projects.

In some cases, it is possible to hire two people: one well-respected leader partnered with a junior clinician to be the technical expert. If no single candidate exists, and the two people can work closely, this tactic can also work.

This is one of the clearest examples of the Psycho 80. Too many organizational leaders only see the "Tech 20" and think "Hey, install the EHR and we're done, right?" Informatics insiders know instead that psycho-socio-political skills account for 80% of a project's success. Those skills are much harder to teach.

The Lesson Learned

Choose informaticists carefully. Your best option is a specialist who is a respected leader. If necessary, choose the leader and train them in tech; don't try to have a techie learn leadership. Or, if you have the bandwidth, bring your psycho-social skills to a partnership with the tech-savvy teammate.

Grow Your Team: Square Peg, Round Hole

How do you keep a bad apple from spoiling the barrel?

An effective leader builds their team to have a wide range of viewpoints. When managed well, the collective wisdom of the crowd is greater than any single brilliant individual. Yet your team may include an individual with ineffective and sometimes destructive behaviors.

For example, a team member who...

- retaliates when their position is challenged
- creates conflict that takes time to resolve
- hinders communication by forcing teammates to tiptoe cautiously
- drives other members out of the team.

What is the source of this behavior? In healthcare, I have never met an individual who has survived both medical school and residency who is out to do bad things. Healthcare is a selfless profession. Everyone has good intent.

How, then, to account for the friction?

Consider the alternative diagnosis: a mismatch in skills. The individual may simply not fit their current position in the organization. Their skills don't match the requirements of the job. A square peg in a round hole.

Individuals with skill mismatches are easy to spot. They draw our attention by causing strong negative emotions like anger, fear, and avoidance. Although they are easy to spot, they are much harder to diagnose and treat.

In my view, there are three options. If the square peg doesn't fit in a round hole, then...

1. Make the square peg rounder

Most people can be taught communication and teamwork. Books like *Crucial Conversations*,[1] *Getting to Yes*,[2] and *Thanks for the Feedback*[3] offer different guidelines for how to talk to one another. Having your team read these books and have a book-club-style discussion may help distill their shared communication, negotiation, and feedback styles and make clear to team members that you value these skills. However, teaching communication, negotiation, and feedback as theory doesn't always translate to changes in behavior or the creation of skillful communicators. The learner must be interested in frequent feedback, practice, and effective coaching.

With feedback, most individuals can improve. Having mentored over 100 individuals over the last 20 years, I believe that most people are coachable. Watching them grow is one of our great joys as leaders.

2. Make the round hole squarer

As a clinician, rank these statements by communication difficulty:

"You never give me enough Percocet for my headaches."

"You don't even listen to me and what I want."

Or, a statement in Hindi from a patient in obvious distress?

Most English speakers would consider the Hindi-speaking patient the most difficult. However, a Hindi-speaking physician would say "That's easy. They're describing classic appendicitis pain." Difficulty is a matter of the skills you bring to the situation.

Using this analogy, a person labeled a "bad apple" has a mismatch between their skills and their job requirements. Mismatches are common. From conversations with leaders at other organizations, I frequently hear about individuals who are not good fits for their groups, or who are not productive employees.

Suppose a team leader has to be unexpectedly absent at the leadership meeting, so they ask their technically proficient employee at the last minute to step in and present a project update. The presentation goes poorly, there are assumptions and misunderstandings, and people raise their voices. The presented project is put at risk as other department heads withdraw support.

The leader could have coached their employee on presentation skills and style. Or the leader could have deferred the presentation until they were able to give it themselves. A good leader would spot the mismatch in skills and make adjustments to the role to help the employee succeed.

At times, we must grow as leaders. If we grow our skills to match the challenges of running a team with diverse attitudes and strengths, then we don't have a problem.

3. Find a square hole for that square peg.

The book *Good to Great*[4] describes the hiring practices at Southwest Airlines. Simply put, Southwest hires by personality and prizes interpersonal skills over technical know-how. Their approach is, "Get the right people on the bus." If you have the luxury of choosing your own team, the Southwest approach is ideal: Choose those with interpersonal skills—technical skills can always be taught later (see **Grow Your Team: Hire the Techie or the Leader?** on page 163).

On the other hand, sometimes you don't pick your own team. Your first task in such a case is to teach communication and collaboration among your teammates. This is hard. If the two approaches described above don't work, then it's time to consider moving the problem employee off your team (and, if necessary, out of the organization). But first:

Get input from other teammates who know this person. Obtain discreet feedback from others. Gather broad feedback to make a better decision.

Set up meetings to share concrete feedback and to set specific goals for improvement. Feedback meetings should be in person. Phone or email or online messaging are insufficient for such crucial conversations. These can be highly emotionally charged meetings, and in-person conversation allows you to use nonverbal channels to reduce miscommunication.

Seek alternative placement for this person's skills. Shuffle the deck. Many individuals can be reassigned to a position that better fits their skills. Perhaps someone is more suited to back-office technical work than customer-facing conflict resolution. Perhaps another person thrives when given team building responsibility instead of technical work.

Set up a concrete timeline and forecast consequences. Once you've diagnosed the mismatch and set goals, the improvement should be evident immediately. Think of how you will address future events. Think through your actions ahead of time.

Keep a careful paper trail of conversations with feedback and goals. Having this trail, whether as email summaries of conversations, or other documents, is important as reference for any future challenges or appeals.

But, there are times when the mismatch cannot be fixed. In this case, it is time to get help with arranging dismissal.

Ask for help from a colleague leader or someone from human resources. If we are lucky, reassigning or dismissing a colleague is a rare event—which means we are not skilled at it. Get help from those with more experience.

Regardless of outcome, it is *worth handling the mismatch.* This shows team members that you have their individual interests at heart. No one wants to work on a mediocre team with internal conflict, and a team looks to its leader to create a safe environment. A leader is more effective when they have the skills to manage team dynamics effectively.

It may be tempting to hope that the conflict is circumstantial, and the problem will solve itself. Remember, internal conflict affects the team as a whole. A **cohesive team** achieves more than individuals can by pulling together and building each other up. A disruptive member ruins that cohesion.

> **The Lesson Learned**
>
> Addressing skill mismatch by individuals on a team is difficult, but it's a critical skill for a leader. Most leaders need help doing this well. Letting the conflict go unaddressed is guaranteed to create long-term consequences.

Grow Your Team: Grow Humans, Not Employees

What do the books *Good to Great*[1] and *How to Raise an Adult*[2] have to do with Informatics? I recently had an enlightening chat with a mentee who was stressed about their slate of recent failures: Medical assistants in the clinic tended to leave their small practice after a year or so of working with this person, and newly hired clinical informaticists under this person's direction were talking about leaving for different jobs. What were they doing wrong?

After further discussion, I also learned that:

- The MAs who worked with this person left to go to nursing school, physician assistant school, and physical therapy school. They were all growing their careers.
- The informaticists who were thinking of leaving were also hoping to pursue bigger and better things.
- Those who *had* left for other jobs often dropped by with a hot lunch or gifts for their former leader--the same leader who was worried about this trend.

So, which is it? Were they a failure or a success?

Initially, when a high-performing individual leaves an effective team, it can feel like a personal blow. "I spent so much time growing and mentoring this person over the past year, and now they're *leaving me*. What am I doing wrong to drive them away?"

In *Good to Great*, Jim Collins talks about "Level 5" leaders who are humble enough to lead from below: to promote teamwork and team members. These leaders celebrate the successes of others rather than claiming credit for themselves. Sometimes, an outstanding employee outgrows their job and leaves for bigger and better things. A Level 5 leader understands and encourages the employee's growth when a lesser leader might selfishly try to prevent them from leaving.

One could choose to see failure when thinking of employees who leave. All that expertise, just walking out the door! Or one could choose to see it as a success: We mentored this person, encouraged this human being to see their own potential, and our shared ideas will be carried into the larger world.

Here's a thought—wouldn't it be a tragedy for a superstar MA to spend a decade being a superstar MA, instead of going on to become a nurse or physician assistant or physician? (Of course, some people will want to stay and *be* that superstar MA, and that is okay too.)

A cynic might point out that an organization is just an edifice, a convenient collection of individuals who temporarily agree to work together for a common goal. The purpose of an organization is to deliver on its strategic goal in a sustainable way. We have all heard of organizations that deliver great products or services while their employees are treated terribly and constantly quitting. Even this can be "sustainable," if there is a ready supply of new hires.

So, no, it is *not* necessary for an organization to care about individuals. Individuals are part of a larger machine. On the other hand, why not aspire to be a next-level organization? In addition to delivering on a strategic goal, why not work toward benefiting individuals as well? "Relationships with management are the top factor in employee's job satisfaction..."

Where possible, a boss should seek to design each employee's job to be "win-win-win." The first win is opportunity for growth. The job should allow your colleague to thrive as a human, find challenges at the edges of their skills, and encourage their personal growth. The second win is strategy. The job should contribute meaningfully to the strategic plan of the organization. The third win is team performance. The job should build relationships that strengthen your team and allows it to achieve big things.

In an interview about her book *How to Raise an Adult*, Julie Lythcott-Haims notes, "...(K)ids aren't bonsai trees. They're wildflowers of an unknown genus and species, and it is my job to provide a nourishing environment..."[3]

Our most important task as a boss is to grow our employees by teaching them effective teamwork, giving them confidence, and letting them spread their wings.

The Lesson Learned

As a leader, your job is to ensure your team achieves its goals. As a leader, your job is also to grow the humans under your care, even if they end up leaving. Your goal is to ensure the organization, the team, and the individual all win. Grow humans, not employees.

INFLUENCE—All the Lessons Learned

Intro—Is It Time to Be the Boss?

Can a CMIO be a CIO? Much of leadership is about building teams and making decisions and carrying them out. Good leaders have common skills and characteristics.

Machiavelli on the Inevitability of Resistance

There will always be detractors and opponents to change. Even good ideas often fail on the first try. When facing defeat, your persistence is what will underlie your success.

Machiavelli on The Grand Gesture

Make small improvements and celebrate them publicly. When you can, aim high and make a "grand gesture" improvement that everyone can feel.

Machiavelli on The Burning Platform

Chosen carefully, a burning platform can facilitate large-scale shifts in behavior. If there isn't a natural crisis, you can create your own.

What Is Your Vision Statement?

Crafting a vision together is a valuable activity. When there are competing demands, a strong vision statement serves as a powerful North Star.

Three-Circle Venn Diagram: A Personal Approach to Influence

Find a passion project that's aligned with system strategy and has a practical path forward. That's a Golden Project with a much greater chance of success.

The Ferret and the Flywheel

Know where your organization stands on the tech adoption curve. This will tell you when to focus on turning the flywheel and when to let your inner ferret out.

Minimum Viable Product: Using the Startup Mentality

The Minimum Viable Product is a good principle for business and informatics. If the big idea isn't working, then try to make it a small idea. Keep It Simple, Stupid.

Subterranean Informatics

Are you hitting an obstacle? Don't go over, go under. Turn on your learning mode. Try a different approach. Skill up. Pivot. Find allies. Re-emerge and try again.

Don't Be Selfish, Be Useful

As an underling, first, be useful. Good things will come. As a leader, reward your direct reports for initiative and good work. You need good ideas from all the minds.

The 70% Rule for Decision-Making

Decision-making and decision-timing are difficult. Take guidance from smart leaders outside of healthcare, and don't wait until you have all the info.

CPOE and the Critical Bug

Leadership decisions are not made in a vacuum. Organizations have different finances, pressures, and cultures. Even the same technical glitch can lead to a different leadership response.

Grow Your Team: Hire the Techie or the Leader?

Choose informaticists carefully. Your best bet is a respected leader with a tech background. If you have to, train a leader in tech—don't try to teach a techie leadership. Or, pair the techie with a leader (yourself, if necessary).

Grow Your Team: Square Peg, Round Hole

Addressing skill mismatch by individuals on a team is a difficult but critical skill for a leader. Get help. Do this well, you create synergy. Do this poorly, chaos ensues.

Grow Your Team: Grow Humans, Not Employees

As a leader, your job is to ensure your team achieves its goals. As a leader, your job is also to grow the humans under your care, even if they end up leaving. The organization, the team, and the individual should all win.

INOSCULATION

Inosculation is the gentle joining of two things, as in tree roots or blood vessels. In this section, we inosculate the field of medicine with the field of information technology, the realm of the patient with the realm of the doctor, and the viewpoints of you and your colleagues. Different fields have different cultures, time pressures, and languages. Where those things intersect, misunderstandings are common.

Intro—Dawn Patrol and Social History

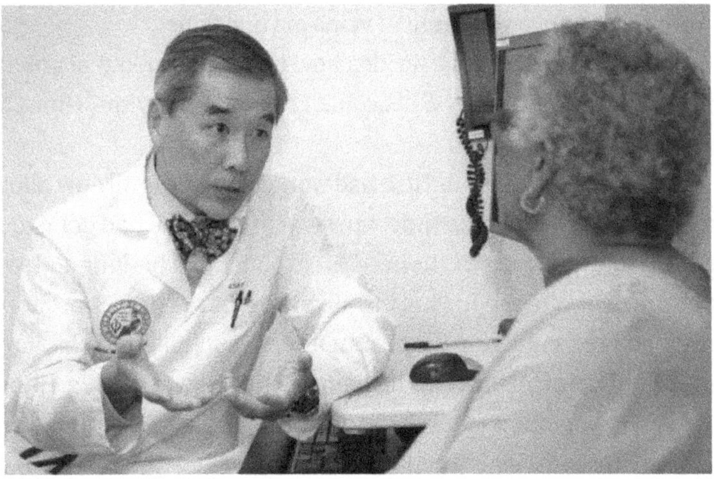

It was a crisp morning in northern California in the fall of 1989, and I was a nervous first-year intern. Our medical team had been summoned to the emergency department. My resident turned to me: "You'll be admitting the patient in Room Three. He's 38 years old. Pneumothorax. Chest tube already placed. Go figure out what happened, write the admission orders, then come talk to me. He might be a good case for Dawn Patrol tomorrow."

Dawn Patrol was a universally feared but educational experience with the Chair of the Department of Medicine. Every morning, the Chair rounded at 6:00 a.m. with the post-call team that had been up all night admitting new patients. Some unlucky intern would gamely present a patient they had "worked up" overnight. The Chair would prod the intern about the patient and their physiology to determine what they knew (not enough), and then some wise words would be offered to the team.

I walked into Room Three. The patient was sitting there in discomfort. A large, heavily bandaged tube pierced the side of his chest and drained to a gently bubbling suction pump huffing quietly in the corner.

I introduced myself. As we spoke, I found that that the patient had unintentionally caused his own problem! He explained to me: "Doc, you know how, when you bite your own thumb and blow, you can look in the mirror and see one of your neck veins pop up, right? It's a great place to put your needle once you're out of veins in your arms."

I thought, "What? I have no idea how it feels to run out of arm veins because of needle scarring. What could I say?" I went with: "Hmm. Sure, I get it."

He showed me how he did it. It had worked well up until now, allowing him to inject cocaine. This time, however, his needle had gone too far. Instead of entering his vein, the needle had punctured his lung and started an air leak. This caused his lung to slowly deflate: a pneumothorax. As more air entered, enlarging the pneumothorax and collapsing his lung, he had gotten short of breath. This happened while he was driving. He began struggling to breathe, and his driving became erratic. Thankfully, he was pulled over by some astute cops who then delivered him to our ED to be treated, and here was, handcuffed to his gurney, in Room Three.

"Great case," I thought to myself. "Perfect for Dawn Patrol." I asked my supervising resident how to prepare. He encouraged me to go back and obtain a thorough history and conduct a detailed examination for the inevitable grilling from the Chair. I read extensively about the physiology of pneumothorax and the chest tube procedure, the anatomy of the neck veins, and their proximity to the top of the lung. I was ready. Or so I thought.

At 6:00 a.m., Dr. Silva walked up to our group of interns. "Good morning! Who's got a case?"

I smiled brightly. "I do, Dr. Silva!" and launched into my polished presentation.

"This is a 38-year-old prisoner and IV drug user, brought in by police for a chief complaint of shortness of breath, and admitted for pneumothorax. He was previously healthy until the morning of admission, when he—"

"Inneresting," interjected Dr. Silva in his distinct drawl. "Pneumothorax. Stop there. What's his educational level?"

"What? Uh..." I had memorized my presentation and was on a roll, but I hadn't foreseen this question.

"His education level."

"Um. He's a... prisoner."

"I see. So, he could be a medical student, and you didn't ask. Perhaps that's how he has access to needles."

"No, sir. But..."

"So, has he recently purchased a truck from the Forest Service?

"...I don't know, sir." My head hung lower.

"Because trucks in our area commonly carry dust from the Central Valley, and he could have coccidiomycosis—Valley Fever. Very common near here. That sometimes results in spontaneous pneumothorax."

"Uh..." I stared at the ground.

"Any travel to the Ohio River valley? Histoplasmosis? Exposure to tuberculosis in an immigration center? Other causes of spontaneous pneumothorax. Is he a rose gardener?"

"I don't know." I began to lose interest in life.

"Because sporotrichosis is a fungal disease commonly acquired by rose gardeners, and it's also associated with spontaneous pneumothorax. So what *do* you know about his social history?"

"He's a prisoner, sir."

"I see. Okay, C.J., let's do this. I'm tired of interns presenting cases and not obtaining an adequate social history, either for diagnosis or for connecting with patients. From now on, presentations at Dawn Patrol will begin with a detailed description of the patient's social history. Are we clear?"

"...Yes, sir." I was too flustered even to correct my own name.

Over the following months, every single intern at UC Davis learned to gather a stellar social history. We knew every patient's educational background, what schools they went to, what they studied and enjoyed, what occupations they held (every one of them since the beginning of time), what hobbies they had, what their families were like, how active they were, what groups they belonged to, and every place they had ever lived or visited. All of it.

Every intern also personally *thanked* me with a punch on my arm for my performance that morning. "Thanks, 'C.J.' Really appreciate it. Social History starts every Dawn Patrol. Great."

It was arduous, but it had benefits. Over the years, my focus on social history influenced my interview style, and my history-taking skills improved. I didn't even notice that I was getting to know my patients better, developing trust between us. I began to see them more as humans and less as diseases.

For instance, I learned one of my patients was a correspondent for National Public Radio. Another used to train with Joe Louis, the heavyweight boxing champ in the 1950s. One of my patients flew in China with the Flying Tigers, the secret American air force unit in WWII. The entire history of the world walked in and out of my exam rooms every day, if I was aware enough to ask about it. Occasionally, one of these clues was also key to unlocking the patient's medical condition.

It is one of my joys to include such gems about the patient in the "Social History" section of the chart and to make sure it displays in my subsequent progress notes, to humanize this person. These random and often amazing connections are some of my favorite moments in medicine.

The Lesson Learned

The social history does three things for physicians—it reveals valuable medical clues, it helps physicians see patients less as "bodies with disease" and more as "humans suffering illness," and a skillfully elicited social history is a powerful way to establish connection. The stronger the physician–patient connection, the more patients are involved in their own care, and the better their medical outcomes.

Similarly, as an informaticist, one way to get patients more involved in their care is to share as much of their medical data you can. In the following chapters I describe my efforts to shift our culture from secrecy to transparency, building EHR tools to encourage sharing information with patients.

The Evolution of OpenNotes

As of 2021 and the Information Blocking Rule, federal law now mandates sharing test results and progress notes with patients. This wasn't always the case.

In 1999, the triad of illegible handwriting, hard-to-transport paper charts, and medical paternalism kept patients in the dark about their own test results and doctor's notes. I felt strongly that this was not right, but I seemed to be alone. My peers believed that secrecy was necessary, that "the medical record is a dangerous place for patients." Slowly, and with many missteps, we built a movement and were able to shift this culture.

The Information Blocking Rule

"I know for a fact that my patients
don't want to see their results before they hear from me."
—Anonymous physician colleague

In 2000, I strongly believed that patients should have more access to their medical information. My colleagues did not agree with me. They thought such access would cause a flood of anxious phone calls and more work for everyone.

That was over 20 years ago, and patient access to their own medical records is now routine. As patients have gained access to their notes and test results, they have shown greater satisfaction with their medical care. Moreover, patient access has *not caused* more work for doctors and nurses.

This change did not come easily.

In the mid-1990s, it was rare for doctors to give patients their own test results. Instead, patients were told, "We will discuss your test results at our next visit," or "if you don't hear from us, there is nothing to worry about." In the era of paper records, clinics inconsistently did the burdensome work of conveying test results to patients (by sending postcards or dedicating a nurse to making hundreds of phone calls per day).

In the early 2000s, as we began to install electronic records, patient portals were so new that most organizations only allowed a rare test result to be posted online, and only after a doctor reviewed the result and deemed it "safe enough" to release. In 2007, our organization was among the first to begin sharing most blood tests and radiology results with patients as soon as they were ready.

However, since only about 10% of those patients had a portal account, we still had to notify most patients by post. The *big innovation* back then was an EHR feature that printed the result with the patient's address so that it was easier to mail in a windowed envelope, thus saving time and reducing addressing errors. (We were so excited!)

The federal Information Blocking Rule was passed as part of the 21st

Century Cures Act in 2021. Now, healthcare organizations are required to immediately release any and all test results to patients upon request.[1] If a patient had a patient portal account, most organizations interpreted the rule as a mandate to release any result to the portal as soon as it was available. Many organizations were unprepared for this change.

The pushback from many physicians, particularly oncologists, was: "Now patients and their families are reading about life-altering diagnoses online *before* I can counsel them about the meaning of the results. This is the cart before the horse.. This is terrible patient care."

One of my colleagues made the statement: "I know for a fact that my patients don't want to see their results before I tell them." This unsubstantiated assertion prompted me and my informatics colleagues to ask, "Is this true?" There was no large-scale study showing whether patients preferred to see their test results immediately. So naturally, we set out to do one.

We (that is, UT Southwestern, Vanderbilt, Beth Israel Deaconess, and UCHealth-Colorado) surveyed over 43,000 of our patients who had recently viewed their test results, with 8,139 respondents (an 18% response rate)[2]. It is the largest analysis of patient portal users' attitudes about seeing test results immediately. Our key findings were as follows:

- Patients who use the portal overwhelmingly preferred (96%) to have immediate access to test results, even prior to discussing the results with their doctor or nurse.
- This preference for immediate release holds true (95%) even among patients who reported that their results were abnormal.
- Among patients receiving abnormal results, 8% indicated that they worried more after receiving their results.

Our takeaway—keep releasing results immediately. Meanwhile, improve the experience for the small percentage of patients who worried more.

The outcome of the study was cathartic for me. By that point, I had been stubbornly fighting for patient information transparency via OpenNotes and OpenResults for 18 years. Learning that 19 of 20 patients preferred immediate release of their results was gratifying and confirmed my long-held suspicions.

(**Update**: It is now May 2026, and several states have written *exemptions* to the federal Information Blocking Rule. A California statute[3] prohibits physicians from immediately releasing results to patients when those results may be sensitive, such as a cancer diagnosis, genetic markers, or HIV. As of this writing, Colorado is proposing a bill to require a delay of three business days for similarly sensitive results.[4]

In my view, these delaying tactics are well intentioned to allow physicians and healthcare teams to notify patients. I believe this approach is outdated, but only time will tell how policies evolve over the coming years.)

The Lesson Learned

Changing the culture in favor of information transparency is not easy. Doctors and nurses tend to believe it will lead to a deluge of telephone calls and messages. Here we have two opposing positions: doctor tradition and patient advocacy. Ultimately, a cultural shift will have to account for both of them—a way of joining the two viewpoints at a place of mutual understanding. This is the core of INOSCULATION.

The Anxiety Problem with Test Results

Since 2007, our organization has given its patients instant access to most test results. I say "most" because we would delay sharing CT, MRI, and PET scans for a week and pathology results for two weeks. Some results could suggest cancer or other serious illnesses, and we wanted to give our physicians time to contact patients individually before releasing the results. This changed with the 2021 federal Information Blocking Rule. Now, *all* test results are released to patients immediately—even the potentially scary ones.

Our 2021 study (see **The Information Blocking Rule** on page 185) established that the vast majority of patients prefer to see their results immediately. However, we also learned that immediate results can be a source of worry for 8% of the patients we studied. How might we do better for these anxious patients?

According to Justin Ross PsyD, a psychologist at the University of Colorado, there are three psychological underpinnings of test-result-related anxiety (sometimes called "scanxiety"). These are Threat, Uncertainty, and Lack of Control.

Threat: Anxiety arises from a real or perceived threat. That anxiety is amplified when the test results indicate a threat to that person's life or health.

Uncertainty: Patients experience three types of uncertainty. They can feel uncertain about the meaning of a medical term, over the timing of discussing the results with their doctor, and about what treatment choices are available to them.

Lack of control: Patients experience lack of control when they receive results at unpredictable times, when they cannot immediately reach their doctors to discuss the results.

Unfortunately, it's impossible to ensure our doctors will always be available for discussion when test results arrive. The next best option is to give patients a framework for interpreting their results. To forecast these concerns, we built a five-step toolkit based on Anticipatory Guidance.

Anticipatory Guidance is well known to pediatricians, who make a point of describing what to expect from growing children: When will they start walking? When will the teenage attitude begin? This helps parents set their expectations and plan ahead. So why can't we advise patients on what to expect when their test results arrive?

Due to immediate release policies, patients will often see their result before a physician can contact them with professional advice and context. Anticipatory Guidance provides some of that context ahead of time.

The Anticipatory Guidance toolkit includes four statements and a question:

1. **I'm ordering a test**. You can see the results immediately or wait to hear from me.
2. **The best case is**..., and this is how I will contact you about it.
3. **The worst case is**..., and this is how I will contact you about it.
4. **Here are resources** you can use to learn more (e.g., Medlineplus.gov).
5. **What questions do you have?**

In the years since 2021, we now advise our physicians and nurses to give Anticipatory Guidance on all complex radiology and pathology orders. Dr. Robert Breeze, a neurosurgeon at the University of Colorado, has always given Anticipatory Guidance when he ordered CT and MRI brain scans. He noted, "The new rules will have no impact on my practice because my patients already know what to expect."

The Lesson Learned

We know that most patients are comfortable with and prefer to see their results immediately. However, some patients will become worried about what comes next. Anticipatory Guidance removes much of that anxiety.

The Naïve Scientist Part 1: "This Is a Bad Idea."

My first big failure was in 2003. I was a young physician-scientist and a member of the cadre of junior physicians at our academic medical center. We prided ourselves on using *data* to make medical decisions. This philosophy is what led me and my colleagues to seek funding from the Commonwealth Fund to learn about the effects of sharing medical records with patients.

My fellow researcher Steve Ross MD and I built a secure web portal called System Providing Patients Access to Records Online (SPPARO, pronounced like the bird). We recruited physicians, nurses, and their patients to study what would happen if we gave patients access to their entire electronic medical record: their doctor's progress notes, their blood tests, radiology, and biopsy results. All of it.

We hypothesized that patients would benefit from having convenient, online access to all their medical information. One possible benefit was to reduce medical error—if patients had access to their own records, they might call attention to incorrect information. Perhaps a patient would spot a missing "penicillin allergy" or a neglected abnormal prostate cancer PSA test that their clinical team had missed. And despite common concern, we thought that this access wouldn't create extra work for their physicians and nurses.

We decided to conduct our study in the academic cardiology clinic. These would be patients with a heart condition under active treatment. We wanted to recruit patients with serious illnesses, not just the "worried well."

It was a challenge to get the seven doctors and their clinical staff to agree to the study. One of the doctors told us, "The only reason I will participate in your study is because it looks like you are doing a rigorous analysis, and at the end of the year, you will know *exactly* how bad an idea this is." We had to work through such strong emotions to arrive at our agreements.

Over the course of the 12 months, we recruited 100 patients in a randomized controlled trial. Half would be the "experimental group" that received access to their entire medical record. The other half were enrolled in the "usual care group," with no access to medical records other than going to clinic visits, receiving mailed letters of test results, and being able to call the clinic nurses with questions or concerns.

We spent the next 12 months measuring what patients viewed and how often they called or sent messages to the nurse. We wondered whether the patients who read their own medical records might get sick less often and require fewer clinic, emergency, or hospital visits. At the end of the year, we met with all the patients and asked about their experiences.

We also asked nurses and doctors about their experiences and if, as they expected, they received lots of phone calls from worried patients. The nurses and doctors had anticipated a flood of patients calling with questions. They were so worried that we agreed to pay a cardiology nurse four hours per week to handle that sudden deluge.

What did we find?

Disappointingly, patients were not healthier (no improvement on symptom scores for heart failure) and did not visit the clinic, ED, or hospital less often.

On the other hand, nurses only spent about five minutes a week answering questions about notes and results. Patients were very appreciative about the transparency. One told us, "I can't get my care anywhere else, no other place shares records like you do." Another said, "I don't always look up my records, but the fact you share this with me means that you don't have anything to hide. I can trust you."

Nurses didn't think the system was burdensome, and they found that patients called with better questions. Instead of patients asking, "What did my echocardiogram show?" They would call and say "I see that the ejection fraction on my echo improved from 22% to 33%. Did my new lisinopril medication do that?" Nurses were surprised and impressed by the sophistication of these questions.

The same person who was against the idea at the beginning said, "This makes me a better doctor. In the past, I might let a test result sit in my folder for a day or two because I was busy with others. Now, because the patient may have already seen the result, I send a message with my interpretation as soon as possible."

In brief, we reported in 2003 what has since been replicated with OpenNotes in 2011 and Information Blocking in 2021: Patients like having access to their own records, and allowing access only makes about five more minutes of work per week for nurses, and has no work impact on doctors.

Despite all that, we faced a final challenge: At the end of the study, the doctors in the cardiology clinic voted to *remove* patient access. "Your project did not improve patient outcomes or symptoms, so why should we continue? Let's turn it off." To them, patient satisfaction, trust, and engagement were "nice to have" but not crucial.

We informed the patients of their doctors' intention to turn off access. In response, we received an outpouring of emails from patients requesting that we keep the system on. With that overwhelming support, the cardiologists reversed course and agreed to keep the system on.

> **The Lesson Learned**
>
> Gather data to support your position. If you can, do a study. However, this is only the first step in changing culture and changing minds.

The Naïve Scientist Part 2: "But, the P-Value!"

I was excited about the success of our SPPARO project. We had demonstrated that it was feasible to show patients their electronic medical records, that patients appreciated it, were better informed, and trusted their physicians more. We had also shown that it did not increase the work for physicians and nurses in clinic. I was eager to take the next step.

We thought the next step was to spread the good news and expand the SPPARO project to *all* the clinics in our health system. This meant 40 clinics, including primary care, medical, and surgical specialties. Feeling hopeful, I presented the SPPARO findings to the Ambulatory Medical Directors, a 40-physician council that comprised the medical leaders of every clinic at the university.

I felt confident as I arrived and set up my presentation. After all, I had the highest quality data: a randomized controlled trial, conducted in our own institution, on our own patients, with statistically significant p-values. Surely, the strength of my evidence would convince my scientifically minded colleagues. I was certain that, soon, all doctors would routinely share test results and notes with patients at the University of Colorado.

I gave my talk, sat down, and waited for the applause.

The first hand raised was that of my friendly colleague in Family Medicine: "This sounds interesting. Please come to my clinic to discuss it further. We might be interested in participating."

A second hand was raised, accompanied by a stern opinion: "I don't know where you went to medical school, but where I trained, the medical record is a dangerous place. This is a terrible idea." The next response was even less charitable: "CT, that is a nice study you did. However, *my* patients are sicker than those straightforward cardiology patients. My patients have more anxiety and psychiatric illness. This will never work in my practice."

Others spoke up: "The medical record is too complicated. It was never intended for patients to read. This will create more work for doctors. No thanks." Pretty soon, dissenting voices murmured around the room, and the table was a sea of head-shaking and sidelong glances. At the vote, most in the room indicated "not interested." My colleagues told me, "We haven't heard of anyone else doing this. Why do we have to be the first ones? We are doing just fine. No, thanks."

The committee chair, initially favorable to my presentation, accepted the majority vote, thanked me for my presentation, and moved on to the next topic. The proposal had failed. I was totally unprepared for the rejection. I gathered my papers, mumbled "thankyouforyourtime," and left.

What had happened? I was certain that my scientific data would carry the day. Instead, I got clobbered by a brick wall. In later years, I would learn a name for this brick wall: "resistance to change." At that time, however, after licking my wounds and doing some soul-searching, I decided my organization was not ready for information transparency, and I set it aside. I gave up.

(Decades later, listening to me recount this story, my informatics colleague Craig Joseph told me he would buy me a t-shirt emblazoned with the words, "But, the p-value!" In those four words, he encapsulated so many of my career failures.)

Having abandoned my crusade for information transparency, I turned my efforts to the demanding work of implementing an electronic health record. This ended up being a decade-long effort, and I left behind my passion and interest in promoting open notes.

Honestly, it was a decade well spent. Deploying an electronic health record changed physician culture and taught my colleagues the value of storing medical information electronically: instant access to charts, simultaneous access by multiple users, and the ability to analyze patterns in a way paper did not allow. It laid the groundwork for physicians using the computer to send and receive messages with other staff and departments. It simplified the process for patients who were already online and could receive results and messages. Clinics loved buying fewer stamps and envelopes to mail test results.

About 10 years later, a competing health system across town started broadcasting TV advertisements: "At our hospital, your health information belongs to *you*." I saw these ads and felt disheartened, knowing we had done the pioneering work a decade prior without meaningful success.

The ads got the attention of our patient communities—and of our physician leaders. The leadership group called me to talk about my transparency project: "Hey, didn't you do research on this a few years ago?"

In my head, my snarky reply was, "I sure effing did. I brought this to you, and you guys stonewalled me!" Out loud, I replied, "Yes, I did. Would you like me to implement that now? With your support, we can make the change right away."

The national conversation on information transparency was also changing. In 2011, the nonprofit organization OpenNotes made news, publishing a national study involving three healthcare organizations, 100 physicians, and 20,000 of their patients. They demonstrated, as we had, that patients valued seeing their doctor's notes, and this transparency did not create more work for doctors or nurses. Their leader, Dr. Tom Delbanco, acknowledged that he was inspired by a poster my colleague and I had made about our SPPARO project. We had presented it years ago, at a national meeting. Dr. Delbanco told me, "If I am the father of OpenNotes, you guys are the grandfathers."

My CIO boss also encouraged me: "Times have changed, and so have you. Maybe it *is* time." I took a big breath, gathered my resolve, and jumped back in. I was surprised at how much I still cared about sharing data with patients.

It turns out that changing minds is not just about p-values and scientific validity. This time, I knew which arena I was playing in: not the "scientific publication" arena, where carefully measured p-values were paramount, but the arena of "leading change." I anticipated inertia and resistance and planned to overcome them by building trust and communicating clearly with both data and stories. This time, I was ready. I was armed with Machiavellian wisdom (see **On Machiavellian Leadership** on page 125):

- **Create the Burning Platform**: I told my colleagues, "Two thirds of patients don't take their medications as prescribed. Why? Lack of understanding. Sharing information with our patients improves their knowledge and engagement. All of our patient- and family-centered care committees have *unanimously* requested access to their medical information. This is the change people want." I also set a firm date: Starting May 1, 2016, our next Epic upgrade, all physician progress notes from that day onward would be shared with patients by default.

- **Create a Guiding Coalition**: Privately, I had met with our CMO council. They had unanimously agreed to support my vision of "information transparency." Then, knowing I had leadership support, I was able to publicly announce: "Anyone wishing to 'opt out' of sharing notes with patients will need to request an exception from the CMO."
- **Create and Celebrate Early Wins**: "Our first seven clinics have been on OpenNotes for six months with 30,000 patients, and they have experienced zero increase in phone calls or other tasks. Patients are viewing their notes and results, and patient satisfaction is high!"
- **Bake the Change into the Culture**: "All progress notes will be automatically released to patients. If you're worried about a patient safety issue, you can hide individual notes, but you cannot change the default."

Thus, in 2016, we flipped the switch and made all notes available to our two million patients. All at once, we became the largest OpenNotes organization in the country as six hospitals, 17 emergency departments, and 350 clinics opened their notes to patients.

I held my breath for the complaints. Instead, as one of my colleagues remarked, "It was a big nuthin'-burger."

Patients were grateful. "I used to bring my sister to take detailed notes at my doctor visits because there was so much information and I couldn't hear it all. Now, I can count on reading the doctor's notes later. I feel like I have the tools to be in charge of my own health."

Our nurses found that patients no longer called asking for their results. In one urology clinic, one third of incoming phone calls had been

"What was the result of my PSA test"? Those calls evaporated as patients looked up their results online.

Years later, I would call this "My Sixteen-Year Journey to OpenNotes." We had gone from a nugget of an idea in 2000 to a full culture shift in 2016.

> **The Lesson Learned**
>
> Even before the federal Information Blocking Rule of 2021, top-performing health systems shared information with patients to improve their care. This is the complex work of Physician/APP Informatics. To get published in medical journals, you need great data. However, to be successful implementing big changes in a large health system, data is not enough. You need different skills to change minds. To change minds, you need Psycho 80 skills, a strong change management plan, and stories to support it.

Fetal Demise Changes an Organization

"I don't think it should take you three days to tell me my baby is dead." I still vividly remember these words. And I remember how the 30 patients and three staff members of the Patient and Family Advisory Council were coldly silent as they looked at me.

A patient—I'll call her Sharon—was looking forward to her third-trimester fetal ultrasound. It was a Friday afternoon. She lay in the dark room as the cold gel and ultrasound probe were applied to her abdomen. When the technologist was done, he said "The doctor will review it and let you know." Sharon cleaned up, dressed, and left to go home.

The technologist completed the images and sent them on to the radiologist. Seeing no fetal heartbeat, they marked the report "fetal demise." The report was automatically sent to the primary obstetrician, whose job it would be to follow up. However, there had been a recent change in hospital policy. Neither the technologist nor the radiologist knew the completed report would *also* be sent to the patient portal, allowing Sharon to access it immediately.

Sharon was driving home with her husband when her phone pinged. She read the report on her smartphone: "Fetal demise." In a terrible shock, she called her doctor's on-call service, but her doctor was unavailable. The on-call doctor was not familiar with the situation. Both were upset about the premature release of a result with such life-shattering implications. Ultimately, Sharon did not connect with her primary OB until Monday, and by then the damage was done; she had suffered through a devastating weekend. Word of this event spread rapidly through the obstetric physician community, causing an uproar. The previous process would have sent the report only to her OB, who would have viewed it on Monday and contacted the patient personally. The obstetrician was trained to deliver such emotionally painful news, but the fetal ultrasound report was never intended to be the way the patient found out.

Within days, I was summoned to an emergency meeting of the OB leaders, during which they demanded a three-day delay in releasing fetal ultrasounds to give the responsible physician time enough to contact the patient themselves.

As a physician leader and an advocate of information sharing with patients, I was looking to balance the needs of the practicing physician with the needs of patients. I was hesitant to add this three-day buffer, as we had been releasing ultrasound results immediately for a decade. And yet, I acknowledged that this was a terrible outcome that should never be repeated.

This crisis occurred in 2017, before the passing of the 21st Century Cures Act of 2021 and its Information Blocking Rule that required immediate release to patients.[1] In 2017, most health systems still delayed delivery of results to patients for days, or even withheld the results entirely. This thinking reflected the paternalistic views of physician leaders at the time, that "patients can't handle the truth without our explaining it to them."

This was a crucial moment. There was clear consensus among the physician leaders in favor of a three-day delay on ultrasound result delivery. And yet, representing patient interests, I did not want to give up on our existing "immediate release" policy.

I had a flash of an idea: Find allies where you can. I could engage our Patient and Family Advisory Council (PFAC), a group of patients serving as the Voice of the Patient for our organization.

I informed the OB leaders that I would take the request to the PFAC. They had always pushed for more information transparency, and I was hopeful that their views on this incident would be helpful.

At the next PFAC meeting, I explained the details of the fetal demise incident to the members, and we had a lively conversation about the pros and cons of immediate release. Near the end of the meeting, one patient stated:

"I don't think it should take you three days to tell me my baby is dead."

There could be no debating that statement.

I took this exact sentence back to the OB steering committee. "This is what our patients are telling us." Doctors in meetings are rarely shocked to silence, but that room was silent.

Then, one of our specialists in high-risk OB spoke up. "We do this already. If there is a fetal demise or other devastating result, we don't let the patient leave. One of us will drop everything—even scrub out of the operating room—to come speak with the patient immediately, in person."

As the meeting concluded, all OBs throughout our organization decided to honor this principle, and all of them changed their routines to adopt the method of immediate, in-person counseling for such a result.

The larger implication is that we did not step back from offering patients immediate release of most test results; over the following decade, we continued to shorten and eliminate time delays on releasing additional results. In the years since, we have kept to the guiding principle of transparency by telling the story of "fetal demise" and remembering how we gave the patient a voice. To Sharon, we are sorry for your loss, and we thank you, as future Sharons will benefit from the lesson you taught us.[2]

And indeed, this story illustrates two useful principles:

1. Make sure that the right stakeholders are at the table. It is easy to overlook your allies. Engaging the PFAC was a crucial move, but not an obvious one.
2. Stories are powerful. Notice that "data" (no bad events for 10 years) does not beat "story" (one woman's terrible experience). Only another "story" ("I don't think it should take you three days...") carried that persuasive power.

The Lesson Learned

This is INOSCULATION—bringing together the viewpoints of patients and physicians. I was faced with a crisis and the demand to batten down the hatches to protect physician workflow and block patients from immediate access to their information, creating a win-lose scenario. Instead, I chose to open a channel and ask for help from the PFAC. Together, we found a win-win agreement, broadened the conversation, and sought a more patient-centered outcome.

A Rising Tide in the MMOLC

World of Warcraft is perhaps the best-known example of a genre of video games called MMORPGs: Massively Multiplayer Online Role-Playing Games. Online computer games entertain us, not only with digital adventure but also by creating enormous human communities to cooperate and compete with. Human behavior is endlessly fascinating.

I wanted to be a gamer, but my terrible reaction time will never come close to that of a teenager. I survived and enjoyed the single-player *StarCraft* campaigns and struggled through *The Last of Us* on easy mode, but the chaos of an MMORPG raid was too fast and intense for me.

However, in the online informatics community at *userweb.epic.com*, I found my people. There, fast-twitch response time was less important than group problem-solving. We shared big ideas, big frustrations, and asked for and gave help to others doing similar work. In other words, it was an MMOLC: a Massively Multi-clinician Online Learning Collaborative. I felt, for once, like an online gamer—a bright flame in a nationwide conflagration of great ideas.

The pinnacle of my experience on *userweb* was navigating the challenge of the Information Blocking Rule. Suddenly, the federal government was telling us to release progress notes and test results to patients immediately, instead of hiding notes or delaying results for days.

Now, any delay in releasing test results could be met with a million-dollar fine.[1] Everyone was struggling to make sense of the rule and determine how to implement it in their organization. The fear was that patients would freak out, or even harm themselves, if they learned about cancer or another deadly illness through this automated release of results.

What would we do?

In commercial healthcare, clinics and organizations often compete to attract patients by delivering better service or showing better medical outcomes. Our healthcare providers operate in a capitalist system and are run as businesses. Consequently, we often hesitate to share our internal best practices in fear of losing our competitive edge.

In contrast, academic medicine is focused on sharing. We share our latest findings so that others can take our ideas and improve on them, advancing our knowledge of science in medicine. The coin of the academic realm is not an actual coin; it is the recognition of important contributions to science. As the cliché goes, "a rising tide raises all boats." In a sense, the userweb served this academic purpose.

Within my own organization, I developed a strategy and wrote education documents on information blocking, including our ideas on Anticipatory Guidance, a way to support patients receiving test results immediately (see **The Anxiety Problem with Test Results** on page 188). I developed a one-page patient handout and a one-page physician/APP education handout explaining the *why*, *what*, and *how* of the Information Blocking Rule, as well as how to adapt to it. I shared this internally with our 4,000 physicians and tens of thousands of clinical staff.

By 2021, many organizations still had not moved toward sharing notes and results. In contrast, our organization had been sharing test results with patients by default since 2007 and notes since 2016, so we had several years' more experience than most.

Recognizing this time advantage, I believed that my expertise and documents might benefit the larger Epic EHR community. I decided to post these documents on *userweb*, where anyone could take them and adapt them for their organization. There was no need for competition and secrecy for something so major: Doing the right thing was the same across all healthcare systems. Float all the boats!

My post became one of the most popular threads on *userweb*, to the point that it affected the performance of the entire site. How gratifying.

I was happy to see the impact of my documents, particularly because the ongoing exchange of ideas on *userweb* in relation to Information Blocking was of great help to me and my own organization. Sharing great ideas can make us *all* more effective.

The Lesson Learned

An MMOLC saved the day for information blocking. Be a flame in the conflagration of great ideas. Unleash your sharing superpower. I hope you find a way to be part of an MMOLC and experience the joy of win-win collaboration. We can INOSCULATE across great distances and share our best solutions.

OurNotes

What comes after OpenNotes? We have shared doctors' notes with our patients since 2016, and our data has shown that patients benefit from it.

What is the next step in engaging patients?

Dr. Tom Delbanco, with the OpenNotes Foundation and Beth Israel Deaconess Medical Center, had an answer: Have patients and doctors coauthor the progress note.

We were delighted to join Dr. Delbanco and the OpenNotes team to build and deploy what came to be known as OurNotes, a tool built on the idea that patients should contribute to the written records of their medical journeys. Doctors are the experts at medical diagnosis and treatment, but patients are the experts of their own history and symptoms. Having both contribute to the medical note seemed like a straightforward win-win.

But how to make it work?

Up to this point, EHRs were intended for use by doctors and nurses. Patients were permitted to send messages and ask questions, but no EHR was designed to let them write in their own medical record.

Creating OurNotes would include technical challenges:

- In our existing EHR, it's easy to get a doctor to write an Interval HPI or a medically focused narrative of the patient's recent history. How do you explain the HPI (history of present illness) to a patient?
- If a patient agrees to write an interval HPI, when should we ask them to write it? How far in advance of a visit? If too far in advance, will the details change? If too near, will patients have time to see and respond to the request?
- What if a patient writes about something urgent, like chest pain or shortness of breath, and no one responds to that questionnaire immediately?

After some brainstorming, we created a set of steps:

1. First, we would ask the patient, "How have you been since your last visit? Have you seen a new provider or noticed a new condi-

tion? Tell us about your symptoms, what you've tried, and any new medications."

2. We would attach this questionnaire to our existing e-Check-in. Patients were already accustomed to verifying their insurance, address, and medication list three days before their visit. Now, we ask them to answer OurNotes questions at that same time.
3. On the questionnaire page, patients were told, "Your provider will see your replies at your visit. If you cancel or do not show up, your replies will not be saved. Do not ask any urgent questions." With this, we have successfully avoided unexpected urgent symptoms from patients.
4. If a patient completed an OurNotes questionnaire, we would automatically insert their answer into the HPI section of the doctor's progress note for the upcoming visit. Then, during the visit, the doctor could edit or delete as needed. The section was titled, "Patient Entered Data: How have you been since your last visit?" This made it clear to the doctor that the words in this section were written by the patient.

In our experience, most patients wrote notes that did not need editing. In this way, both patient and doctor contributed the data from that visit, improving communication and collaboration. The progress note became co-authored, with the patient contributing history and symptoms, and the doctor contributing objective findings and medical decision-making.

Beginning in late 2019, we tested OurNotes at two clinics. Over 12 months, we showed that more than 90% of physicians and APPs viewed OurNotes positively and wanted to continue using it, as did over 90% of patients.[1] Doctors felt that OurNotes saved them time spent writing and included more of the patient's perspective.

Our pilot test, conducted with Beth Israel Deaconess Medical Center (BIDMC), was so successful that we launched OurNotes for all 70 primary care clinics in our organization. About half of our specialty clinics have also joined the OurNotes process, with patients writing about their medical histories prior to seeing their specialists.

About 20% of patients write in OurNotes prior to their appointments, and physicians/APPs use the notes regularly. Not all patients respond to the OurNotes questionnaire ahead of their visit. Those who don't respond have a regular doctor visit, unchanged by the OurNotes process.

Our organization sees about four million patient visits a year, and several hundred thousand OurNotes are written by our patients and incorporated into our shared medical records annually.

The OurNotes trend is expanding. I am aware that at least Sanford Health and UCLA have joined the original OurNotes original research partners: BIDMC, University of Washington, Dartmouth-Hitchock and UCHealth-Colorado. Thank you to all the trailblazers out there for transforming patient-centered care.

The Lesson Learned

Did you spot both INOSCULATIONS? Great ideas often come from outside of your own health system. OurNotes is an example of borrowing great ideas from smart colleagues. There are so many great ideas out there.

OurNotes also successfully combines patient and physician viewpoints. Working together, our innovations can change healthcare.

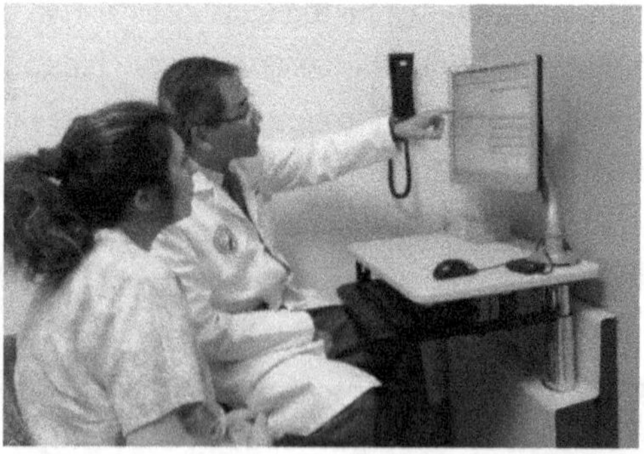

With OurNotes, the medical record is a win-win collaboration

The Cloud Inside the Triangle

Large health systems are built around predictability. A CEO at the top supervises a cascade of leaders so that daily work can go smoothly. This structure is efficient and predictable for "keeping the lights on" and performing standard work, but it can interfere with new ideas and rapid decisions.

Informatics teams, on the other hand, are more like startup companies, with a leader and a small team of skilled individuals. Small teams are nimble and make quick decisions. But as teams get larger, such fluid decision-making becomes unwieldy and chaotic.

Large organizations may thus be imagined as a triangle with a leader at the top, whereas small startups may be thought of as a cloud, with the leader in the center. These shapes don't clearly fit together. How can a nimble, cloud-shaped startup exist inside a structured, triangle-shaped organization?

More to the point, how might an informatics team managing a rapidly evolving EHR thrive inside a large health system that values predictability?

At first, I thought this problem would be fun and easy to solve. In 2011, I hired my first four physician colleagues to be informaticists. In the early years, there was a lot to work on, and we would make decisions in our small group. We built and deployed quick projects within our EHR: new grouped orders to reduce error, innovative alerts, disappearing help text that showed up in just the right spot—a volley of improvements. As we saw success, we applied our changes throughout the entire organization. We were making a difference!

Our team sometimes bypassed committee approval. "Haha! You can't catch us!" We were physician builders with insider privileges. We could build our own tools without submitting to the usual slow process of governance committees. We were fast, we were innovative, we got stuff done.

Soon, however, our nimble changes started to cause consequences.

Move Fast and Break Things: A handful of hospitalists had a specific progress note problem. One of my informaticists built a speech-enabled note that these clinicians could use to dictate a quick paragraph on their smartphones. Then, with a keystroke, they could deftly insert that paragraph into a more comprehensive progress note on the workstation. The original dictated paragraph, no longer needed, would disappear. Those who knew how to use it seemed like magicians.

A couple years later, other users stumbled across this note type, and not knowing how it worked, started to write notes that "disappeared." We were brought to the architect's meeting and asked, "Who said you could do this? Did you clear this with the governance group to avoid these unintended consequences?" Hmm. We had not. The saying in startup culture to "Move fast and break things" might not be a great idea in healthcare.

Our ultimate solution? We created matrix relationships where an individual reports to more than one leader, and the workload is carefully balanced.[1] You might call this a sophisticated inosculation. My informaticists were already clinicians as well as information technology experts: one level of inosculation, gentle interconnection. Now, they also had to serve on governance committees (order sets, documentation, HIM health information management, and scanning decision teams). This was all a second level of inosculation between informatics and the IT teams.

These crossover members are crucial; they link the rapid innovation of the Cloud and the predictability of the Triangle. These members are chosen carefully—they should be able to communicate effectively between their teams and craft win-win scenarios. The Triangle team exists to understand and support what the Cloud is trying to do; the Cloud team benefits from Triangle team experience and historical knowledge.

For example, specialist physicians come requesting an EHR tool to improve a specialty process—usually an interruptive alert and a daily report of patients. Instead of just saying "yes" or "no," informaticists will ask detailed questions about the problem. Often, they can present an simpler, less intrusive solution.

"I think you want a *silent* alert that just notifies a clerk and builds a list. Then you'll interrupt the doctor far less often."

"Oh. I didn't know you could do that."

This model has also been applied to the EHR Optimization Sprint team (see **Sprint Informatics** on page 105). The team comprises a 22-member Cloud of physician and nurse informaticists working alongside analysts and trainers who are part of the Triangle.

The Sprint team does nimble work: fixing broken tools and building new ones within hours or days. At that same time, these informaticists and analysts are deeply connected to existing teams and governing bodies in the Triangle. The relevant naming conventions, IT build principles, and change control rules are all followed; no rules are broken. Our users are happy with the quick fixes, and our governance and leaders are happy that rogue agents aren't causing chaos.

We employed a similar structure when establishing our "command center" during the COVID pandemic.[2] A command center brings together leaders from all disciplines for a daily huddle. The larger triangle stayed in place, with a nimble cloud assembling for an hour each day. We were able to solve pandemic issues in rapid succession, addressing daily changes in isolation and ventilation policy. One ICU doctor taught us to pronate the sickest COVID ICU patients, and we taught this best practice to all clinical teams across 12 hospitals within a single day. We retrained several hundred idled clinic nurses and doctors, allowing them to assist the overwhelmed inpatient services. We trained 2,000 doctors and nurses to use telehealth to see patients remotely when in-person care became impossible. These wins were only possible with a nimble team supported by a strong outer structure.

> **The Lesson Learned**
>
> Fostering a startup culture in a larger organization is difficult. It requires an effective leader who can build trust and accountability and communicate well. Health systems *must* learn to innovate and

be nimble while operating smoothly. Creating "a cloud inside a triangle" with matrixed relationships is a strong solution.

Tribes in Medicine and Building Successful Teams

Why do some teams outperform others?

The book *Tribal Leadership*, by Dave Logan,[1] describes five levels of human tribes. All organizations, he argues, are built up of tribes, and each of those tribes has a dominant culture. This culture represents how each member views themselves and their fellow tribe members. The spectrum of views can be categorized into five stages, in which a stage-one tribe will always be outperformed by a stage-two tribe, and so on.

To paraphrase:

Stage I Tribe: "Life sucks."

Stage II Tribe: "My life sucks."

Stage III Tribe: "I'm great, but my colleagues suck."

Stage IV Tribe: "We're all great."

Stage V Tribe: "We're world-class."

Over the years, I have been in many working groups. I have managed to avoid Stage I Tribes, but I believe I have seen every other tribe in my years as a practicing physician and informaticist.

Surprisingly, I know quite a few physicians who are in Tribe II: "my life sucks." Many have survived punitive or toxic cultures in their years of training, and they reflect these toxic attitudes in our meetings. If they are not succeeding, they'll make sure you don't either. Comments from colleagues living in Tribe II: "That is a terrible idea. What were you thinking?" "Why can't you just get rid of this stupid policy?" "Leave me alone. You can't help me." Self-doubt, anger, and counter-productive behaviors characterize these Tribe II members, who seek to drag others down to their level.

More enlightened colleagues may be in Tribe III: "I'm great, but my colleagues suck." This is the most common tribal culture (apparently affecting about half of all American employees). These individuals complain constantly that "I'd be able to get sh*t done, if it wasn't for these idiots I work with." Many physicians I know believe that they alone provide outstanding patient care, and everyone else stands in their way. Imagine a character like House, MD: "Thanks to that individualistic, brilliant physician who didn't trust anyone else and repeated the entire work-up themselves, this patient was diagnosed and cured."

In working on committees with physician colleagues, I have seen plenty of Tribe II and Tribe III behavior. It makes for a challenging meeting when you have to manage egos and gently redirect those individuals to the larger goals of the group. Many Tribe III physicians never received leadership or teamwork training. They put their teams at risk. Physicians with Tribe III attitudes ("You'd all be lost without me!") who serve as committee members or leaders do not foster collaboration. Their main tactic is to push their own ideas and force others out of the way.

With coaching, an individual can rise from one tribe to the next, but it takes persistent nudging over time. It is *not possible* for someone to skip a level, say from Tribe II to Tribe IV. I once tried to push a mentee from Tribe II ("my life sucks") up to a Tribe IV mindset. Do not try this. It was confusing and everyone was unhappy. Instead, I had to coach them to realize "I'm great" and be happy that they reached Tribe III. It would be years before I was able to nudge them toward Tribe IV.

Occasionally, a group forms that reaches Tribe IV: "We're great." Members of this team see the bigger picture and recognize that their team members have valuable viewpoints. Further, they realize that the team can accomplish more than any individual could.

I like to think we have built several such groups. Our Sprint teams for EHR optimization, our "Large PIG" physician/APP informatics group (see **What Is Your Vision Statement?** on page 137), and a growing number of working groups led by our Large PIG members.

Tribe IV members grow to know and trust each other over time. When the conditions are right, Tribe IV behaviors emerge: They consistently communicate, both verbally and in writing, in a way that supports their team members. There are group norms that gently correct egocentric behavior and support the quieter members of the group.

While meeting with our Electronic Medication Optimization (EMO) committee, I realized we had built such a Tribe IV culture over the last couple years. The conditions were right: we had about seven members, all with complementary skills: IT analysts, clinical nurse informaticists, a pharmacy informaticist, and physician informaticists, each with a differing viewpoint but the shared goal of squashing low-value alerts in medication ordering.

In our group meetings, I had noticed the easy banter, the proposal of ideas, and the very different viewpoints expressed. Even more, I enjoyed coming to a gradual consensus to turn off low-value alerts despite years of historical tradition and personal discomfort toward change and uncertainty.

How does this happen? I consistently hear tentative, vulnerable statements from all members, including the leader: "This might be a terrible idea, but..." or, "I'm curious what you think of this..." instead of, "Here's what we need to do." There is explicit turn-taking. No one speaks over anyone else. This is particularly noticeable during online meetings, when the audio and video latency cause two people to start speaking at once. There are instant apologies and a negotiation as to who speaks first. Almost always, the more junior person is encouraged to start. Team members actively solicit the input of quieter members.

As I see each of our members listen to the team and shape a careful decision, balancing patient safety, prescriber efficiency, and reducing unintended consequences, I notice the changing of minds and a humility of "Oh, I see what you mean." These words are only spoken in meetings where participants feel safe.

These behaviors are easy when membership in the committee is stable and members have implicitly agreed on how to work together. When new members arrive, it is important that the leader(s) enforce the meeting norms with the new person: "Hold on, Joe, let Sarah finish her thought." Or perhaps the leader speaks with Joe after the meeting for immediate feedback. Lacking such correction, a Tribe IV meeting could easily descend to III, with less safety, vulnerability, and effectiveness.

On the other hand, when a Tribe IV team continues to work on relationships and teamwork skills, there are moments when it can achieve Tribe V and make remarkable breakthroughs. Stage V Tribes are non-sustainable cultures that can exist only when many factors are perfectly aligned. A consistently high-performing Tribe IV team may occasionally emerge into V, and this is the pattern we strive for.

In these quiet moments of hard thinking and intense collective work, I enjoy sitting back and seeing the gears turning in everyone's heads. And then I end up blurting out how happy I am. "We are a stage IV tribe, y'all!" I would wish for all my teams to consistently reach stage IV.

My teammates, shocked, looked at me like I'd grown a second head. Then I had to explain myself and the entire concept of Tribal Leadership, stealing precious time from our meeting. However, once things are explained, there are usually smiles. People are often working so hard, focusing so intently, that they don't notice how well we mesh as a team. Calling attention when things go well is a joyful responsibility of a Tribe IV leader.

What does this sound like? I have heard myself say things like:

"Guys, I just want you to know, I love working on this team."

"Can I just say, this is the best part of my day, working with you guys on this."

You don't need to be a team leader to point out your team's cohesion. When you notice people working well together, mention it! If you ever find yourself smiling broadly at how fun and effective your team is, say something. This can lead to comments like: "Hey, let's do that again!"

The Lesson Learned

Joyful, well-run teams are rewarding, effective, and self-reinforcing. I think of *Monsters Inc.*, the Pixar movie from 2001 in which capturing "scream energy" from frightened humans ended up being less potent than capturing the "laughter energy" of joyful ones. I have found that taking 10 to 30 seconds to appreciate the great work of the team or individuals on the team is powerful and leads to more expressions of gratitude, more selfless behavior, and more creativity toward the goals of the team. Catch your team in a Tribe IV moment and inspire them to make more moments like that.

"The More I Listen, the More My Patients Learn"

The essence of inosculation is bringing together people with opposing viewpoints. Sometimes those people are patients and physicians who view OpenNotes and Open Results differently. Sometimes, they are committee members with different values. Could a communication technique bring us together?

Two decades ago, after a few years seeing patients in a busy internal medicine practice, I had burned out. Sure, I was great at the internal medicine stuff: diabetes, blood pressure, heart, lung, stomach problems, and more. But I was tired of the endless parade of vague symptoms and self-destructive behaviors (lack of exercise, alcohol, cigarettes) for which my medical knowledge was ineffective. I felt powerless. Why wouldn't patients just make a simple change when I told them how to fix their own problems by changing their lifestyles?

In the late 1990s, the entire junior faculty at the University of Colorado was invited to attend a four-hour communication workshop held by the Bayer Institute (later The Institute for Healthcare Communication).

At this workshop, I learned communication tools and found an entirely new way of thinking about patient care—one that emphasized demonstrable caring and relationship building. And I came away with not just theory but practical words I could use.

I was excited enough that I spoke to my workshop instructor, Dr. Fred Platt, after the course and asked if I could learn to teach the material as a way of understanding it better myself. To my surprise, he took me under his wing.

I traveled to West Haven, Connecticut to learn to facilitate the physician–patient communications workshop. I met 30 other physicians sent by their respective organizations. I felt like I belonged to an elite club!

The next day at the training center, my big puffy head and I were not at all prepared for the intensity of the interview. I would be working with a "standardized patient," an actor mimicking a common patient scenario.

Before we began, I was told, "You are a surgeon going to see your first patient of the clinic morning." I walked in, sat down, and said, "Good morning, Ms. Smith. How can I help you today?"

The patient screwed up her face in anger and... threw her plastic water bottle at me! It bounced off my chest as she yelled, "You have f***ing ruined my life! *And you don't even care*!"

Eyes wide, I froze. After what felt like minutes of my terrified silence, the coach sitting at my elbow signaled a time-out. "Maybe you'd like to pause here? What is going through your mind right now?"

What was going through my mind was, "*Oh my god*. I have to get out of here!"

Eventually I calmed down, we debriefed the situation, and I decided to try it again. It took several repeated attempts and constant at-my-elbow coaching before I could even start a conversation with my extremely upset patient.

In hindsight, it was a perfect example of "limbic hijacking." The limbic system in the brain regulates emotion. Limbic hijacking describes a moment when emotions run so high that the heartrate spikes and the reptilian "fight or flight" hindbrain takes over. The moment she threw her water bottle at me, my poor prefrontal cortex (PFC) forgot every problem-solving and communication skill it had ever learned. Then it got rudely shoved aside by my protective reptile brain. I was hijacked.

Me: Help! Prefrontal cortex? What constructive words can I say here? Quickly!

PFC: [petrified silence]

Hindbrain: I know! Pretend you're getting paged! Get up and run away!

Me: Great. A lot of help you both are.

Over the next week, we practiced many strategies to build a connection between doctor and patient, until I could survive that hostile interaction.

Excellent communication is at the root of so many great things. One of the patient interview frameworks we learned was E4: Engage, Empathize, Educate, and Enlist. The E4 framework is a dramatic contrast to the way many physicians interview their patients, which could be called F2, or "Find It—Fix It." F2 is a singular focus on medical data. Because of F2, a physician often ignores the patient's ideas, feelings, and values. Understanding these elements of a patient's viewpoint is crucial in building a strong relationship between doctor and patient.

The E4 framework could be summed up in a quote from a physician colleague who told me: "The more I listen, the more my patients learn." This beautiful, seemingly contradictory statement has a lot to teach.

Reflective Listening and Advanced Empathy with Patients

Has a patient ever told you something like this?

"Hey doc, those pills you prescribed didn't work. I'm still gaining weight, and my blood pressure is high. Maybe you should give me something that works for a meat-and-potatoes guy like me. I like my prime rib. And I hate when everyone lectures me about exercise. I just want to see my daughter graduate."

This statement cries out for "empathize," the second skill in the E4 framework. Empathy includes reflective listening, which is active listening, nonverbal eye-contact and head-nodding, and responding with not only medical data but *ideas*, *feelings*, and *values*. This helps the patient feel their concerns are truly being heard.

Potentially useful reflective statements:

"It sounds like your blood pressure is still high." This is Data. Most physicians imagine an exchange like this, responding only to medical data, when they think of reflective listening. This is a powerful technique on its own. However, consider taking this skill to the next level to achieve Advanced Empathy by adding in ideas, feelings, and values.

"It sounds like you are a meat-and-potatoes guy." (An idea.)

"It sounds like you hate getting the exercise lecture." (A feeling.)

"It sounds like you want to see your daughter's graduation." (A value.)

If you do this well, you will hear the patient respond, "That's right!"

Offering statements about the patient's ideas, feelings, and values is a concrete way to build trust. Before, I thought these statements were just "fluff" and not part of being a medical doctor. I ignored them and focused on symptoms and data like blood pressure and weight. Whoops.

In a medical interview, many things are happening: A patient is telling a story, a doctor is listening for medical clues. More than that, the doctor must pay attention to everything that is said, and *how* it is said, and work to build a relationship. It seems like too much. With practice, however, it becomes easier.

Once you verbally acknowledge the patient's ideas, feelings, and values, they become part of the conversation. The patient feels heard, and there is a chance to build trust. From trust might come openness to listening and maybe even a willingness to change behavior.

In the wise words of Yoda: "Fear leads to anger. Anger leads to hate. Hate leads to suffering."

To paraphrase Yoda for medical interviews: "Listening leads to understanding. Understanding leads to trust. Trust leads to behavior change."

After I'd learned 10,000 medical terms, studied dozens of textbooks, and interviewed thousands of patients, this new framework came as a revelation. The seemingly simple chat between doctor and patient was so much more complex and interesting than I previously knew. It was, yes, a logical search for symptoms and signs of disease, but it contained the more complicated task of building a human connection.

With this revelation, I felt recharged! I could enjoy patient care again.

Reflective Listening and Buckets in Meetings

In the years since West Haven, I have found many other places to apply these communication skills. In fact, we are so frenetic at work that you regularly hear people talking past each other, trying to get their own points across.

Have you heard these phrases in meetings?

"Like I said before..."

"Again..."

"I don't see why..."

"You don't get it."

Every one of these sentences illustrates a mindset of trying to convince others, not a mindset of listening or seeking to understand.

In meetings, I listen for well thought out points. I listen for crosstalk and misunderstandings and untangle them in my head as the meeting goes on. I listen for participants talking past each other and not listening to others' positions. Importantly, participants rarely summarize the statements of others, *especially those with whom they disagree.* Consequently, no one feels heard. People start repeating themselves, the meeting gets chaotic, and sometimes there are raised voices. That's when I raise my hand and step in.

I say something like: "I have heard three *buckets of conversation* so far. Bucket One is... Bucket Two is... and Bucket Three is..." I call this my "buckets" summary. One by one, I summarize what I have heard from all viewpoints, including viewpoints opposed to my own. This last task is difficult to do well. Everyone must feel heard.

If I can, I try to reset the discussion. "Can we agree that for Bucket One, we don't have the right participants, so we can defer until they are present next time? For Bucket Two, can we agree that participant A has a straightforward idea. I did not hear any objections. The big discussion is Bucket Three. Here's my understanding of that issue. Do we agree to spend the rest of our time solving Bucket Three today? If so, may I propose we try X? What do we think of this?"

I find that after a strong summary statement, if all the parties feel heard, the cross-chatter stops. The remaining minutes can then be employed toward solving a more clearly defined problem.

There are some caveats to being a bucket summarizer. You have to choose your timing carefully. It may not be my place to jump in on another leader's topic, for example. Am I strongly identified with one of the opposing sides in a discussion, so that anything I say is already judged against me? Perhaps this is not the time to speak up.

Sometimes, meetings already have a good leader or a skilled participant who will actively steer the conversation toward an effective resolution. At those times, I am happy to keep quiet. If it is my team member, I am joyful for their success, and give them kudos afterward for a productive intervention.

> **The Lesson Learned**
>
> When speaking with patients, restrict yourself to talking less than half the time. Focus on reflective listening and ask to hear more. Summarize the medical data, as well as ideas, feelings, and values. This is difficult at first, but it is highly rewarding for both patient and doctor. Done well, it does not take more time. "The more I listen, the more my patients learn." With patients, listening is INOSCULATION.
> In a meeting, see if you can listen well enough to serve as a summarizer. Or, listen for someone who does this well, and learn from their example. It is wrong to assume that disagreements in meetings must result in a "winner" and a "loser." It is better to assume there is a possible win-win outcome. We just need to listen hard enough and be clever enough to find the solution. With colleagues, also, listening is INOSCULATION.
> Or, to paraphrase the original quote: "The more I listen, the more my colleagues agree."

(For more reading on these skills, see **Appendix 1**.)

Useful Information Hides Inside a Feeling

In my own head, I am the only reasonable person on this planet. Everyone else is a dummy. Have you read Fredrik Backman's *A Man Called Ove*[1] (or the corresponding movie, *A Man Called Otto*)? In it, Ove is a grumpy old man who thinks everyone else in the world is an idiot. And, against your better judgment as a reader, the author makes you come to love this man, his struggles, and his viewpoint.

I am easily derailed by feelings and emotions. Recently, I had this thought: "Why does this person contact me at the last minute, and expect me to lead the discussion on an unfamiliar topic at the meeting tomorrow? What an idiot." From there, I fired off an email about their lack of planning. And then regretted it.

One of my classmates in medical school coined the term "limbic hijack." The limbic system in the amygdala of the brain reacts automatically to strong emotion, triggering "fight or flight" in response. When the limbic system is activated, the thoughtful frontal cortex is bypassed, and you take action as if your life were in danger. In the wilds of the Serengeti, a great idea. In the modern office, not so much. I had fallen prey to a limbic hijack and sent a quick email without a lot of thought.

When my recipient read the email, they picked up the phone and called me. "Hi! I'm surprised to get this reply from you. What's going on?"

Our misunderstanding was resolved with a five-minute phone conversation. What happened there?

1. Words on a page are a terrible medium for communicating emotion (not good).
2. I had gotten hijacked and let my emotions lead (not good) instead of pausing to reflect.
3. My colleague detected my frustration and escalated to a phone call (good!)

Being aware of your own feelings is an excellent diagnostic skill, in both the exam room and the board room. Practice every time you feel a strong emotion: Categorize it and pause before you act.

It is important to listen to your feelings, because they often clarify complexity.

I was recently mentoring a colleague who was struggling with a decision. I described a Gedanken experiment, laying out three very different futures, and asked my mentee to spend an hour or two writing about each future, imagining themselves in it. Often, when I suggest this, my mentee will do the exercise and then call to tell me "Out of the three scenarios, one of them jumped out at me, and I got really excited."

Gut feelings are clues that can cut through the haze of conscious analysis. For instance, we may be in a situation that has too many moving parts, involving too many people with different agendas, to make sense of the situation and come to a rational decision. We are stuck with too much data and no clear direction. In these difficult cases, if we have studied deeply, it is best to step back for some time and let your subconscious mind work. Does a feeling emerge? Do you have a feeling or sense that one direction is the better choice? Do you have a "bad feeling" about a certain choice? Feelings, in this way, can be a helpful guide.

Keep in mind, what I am calling "gut feeling" is not the same as betting on horses at the racetrack based merely on the name of the horse, your hunches, or a "hot tip" from a stranger. Instead, these gut feelings may emerge after you spend time getting muddy, diving deeply into the facts and doing your homework studying the situation. Daniel Kahneman, in *Thinking, Fast and Slow*,[2] writes that we should mistrust intuition and "gut feelings" unless three conditions are met:

A. In the area of study, there are patterns your brain can pick up on. (In other words, there is something meaningful to analyze.)
B. You have a lot of practice in this area.
C. You receive immediate feedback in this area.

If the choice you're making is not intellectually clear *and* the three conditions above are met, in those situations, you *could* "go with your gut."

Noticing Reflective Emotion

Our bodies easily mirror the feelings and nonverbal appearances of others. Two people in an enthusiastic conversation may gradually get louder and speak more quickly to match each other. Another two people may both cross their arms when disagreeing.

When working with patients, I use my body's response as a diagnostic tool. I find that pausing to look at my own body language and my feelings can give me insight into my patient. If I look down and find that my fists are clenched, I can pause and recognize that I'm frustrated or angry. What am I reacting to?

If I'm feeling trapped and sad just listening to a patient in the exam room, that tells me something about what the patient might be going through. I also find that bringing up my own feelings allows a strong connection to my patient and sometimes leads to a breakthrough in our discussion:

"Joe, I'm feeling angry and frustrated when you tell me about your family. If I'm feeling this just as I listen to you, maybe this is bothering you more than you're telling me." Acknowledging my own feeling can sometimes lead to a "shock of recognition" in my patient that changes the direction of our medical interview. Calling attention to emotion may lead to an insight.

What applies to patients also applies to colleagues. Calling attention to my emotions, asking whether my colleague is feeling this, can lead to an insight. "I'm frustrated with our lack of progress on this topic. How are you feeling about this?" This could prompt a different approach.

Feelings Are Critical in Design Thinking

There is an innovation approach called "design thinking." Its key elements are: Gathering Data, Seeking Opportunity, Mining for Insight, and Prototyping a Solution. The design thinking emphasizes attention to emotions. During the ethnography phase, you will interview a person experiencing a problem you're hoping to solve. In that data-gathering interview, listen hard for expressions of joy, anger, sadness, and confusion. They are telling you what is important to them, and what most needs fixing. Subsequently, when considering possible solutions, the key is often the information surrounding a strong emotion that leads to a breakthrough design.

As informaticists and technology specialists, we are often immersed in in STEM-centric left-brain thinking. Conversely, creative, intuitive, and holistic thinking are influenced heavily by right-brain emotion. In our work, it is easy to become blind to this rich source of insight.

The Lesson Learned

Feelings from the right brain have a place alongside left-brain logical thinking. Beware of limbic hijacking in your own work. You aren't your feelings, and your feelings are not you. Don't be consumed by them. Use them to diagnose the situation.

Consider your emotions if your intellectual decision-making is stumped by complexity. Sometimes a feeling can neatly summarize an otherwise paralyzing amount of data. Similarly, detecting and mirroring feelings in others can lead to an insight. Use your own feelings to tentatively diagnose how others might be feeling and use that knowledge.

INOSCULATION—All the Lessons Learned

Intro—Dawn Patrol and Social History

Social history is surprisingly effective for both human connection and medical diagnosis. Similarly, information sharing inosculates the fields of IT and medicine. In both cases, sharing improves the care of patients.

The Information Blocking Rule

Changing the culture in favor of information transparency is not easy. Here we have two opposing positions: doctor tradition and patient advocacy. Ultimately, a cultural shift will have to account for both of them—a way of joining the two viewpoints at a place of mutual understanding. This is the core of INOSCULATION.

The Anxiety Problem with Test Results

Some patients are more anxious when their test results are delivered without context and their doctors are unreachable. Preemptive Anticipatory Guidance helps reduce this anxiety and gives patients a framework for responding to their results.

The Naïve Scientist Part 1: "This Is a Bad Idea."

Doing a study is difficult, but it is also rewarding, even on a small scale. Data-gathering is the first step in any cultural shift—you need data to be taken seriously. If you can, do a study. However, this is only the first step in changing culture and changing minds.

The Naïve Scientist Part 2: "But, the P-Value!"

...However, data isn't everything when it comes to changing minds. P-values are great for scientific research; they are insufficient for inspiring change. Learn the Psycho 80 skills and build a strong change management plan.

Fetal Demise Changes an Organization

Where possible, inosculate patient and physician viewpoints. Seek allies. Remember that in many cases, a good story beats ironclad data.

A Rising Tide in the MMOLC

Our capitalist culture encourages us to hoard secrets, but in medicine and in informatics, sharing your best ideas pays dividends. Be a bright flame in a nationwide conflagration of great ideas.

OurNotes

Did you spot both INOSCULATIONS? 1. Great ideas often come from outside of your own health system. OurNotes is an example of borrowing great ideas from smart colleagues. 2. OurNotes also combines patient and physician viewpoints. Working together, our innovations can change healthcare.

The Cloud Inside the Triangle

Fostering a cloud-shaped startup culture inside a larger triangle-shaped organization is difficult. It requires an effective leader who can build trust and communicate well. Health systems must learn to innovate and be nimble while operating smoothly. Creating "a cloud inside a triangle" with matrixed relationships is a strong solution.

Tribes in Medicine and Building Successful Teams

Joyful, well-run teams are rewarding, effective, and self-reinforcing. Consider the power of laughter energy versus the power of scream energy. Then, go catch your team in a Tribe IV moment and go make more.

"The More I Listen, the More My Patients Learn"

Reflective listening improves patient care and streamlines meetings. Listen and summarize not only data, but also ideas, feelings, and values! Use "buckets" to clarify discussion. Remember that "The more I listen, the more my colleagues agree."

Useful Information Hides Inside a Feeling

Right-brain feelings complement left-brain logic. Beware of limbic hijacking. A feeling can summarize complex data. Mirrored feelings may lead to insight. Feelings lead to design thinking insights. Useful information hides inside a feeling.

THRIVE

As your skills and knowledge grow, you will be asked to do more, to manage your time more effectively, to contribute to more projects. What self-management skills should a CMIO cultivate? How can you set yourself up to thrive?

Intro—The Eppendorf Tube and the Boston Winter

The first snowflakes had begun falling at dusk, and I was again alone in the spotless, brightly lit research lab. I had already logged 20 weekend hours on this multiday experiment. With still more hours to go.

As a college junior majoring in molecular biology, I was considering a scientific research career. Gathering data for my undergraduate thesis, I was studying P4 bacteriophage, a virus that attacks *E. coli* bacteria. By spending weekends in the lab, I hoped to prove (to others and to myself) that I had what it took to be a good scientist.

The tasks for this experiment, I could still now perform from memory:

1. Take a frozen sample of virus out of the subzero refrigerator, warm it up, dip the Q-tip into the tube. Twirl the Q-tip into a one-liter flask of *E. coli* bacteria.
2. Place the flask into a 37°C rotating incubator for an 18-hour growth cycle. The virus would produce millions of copies overnight.
3. Spin the resulting culture so that the DNA settles to the bottom.
4. Using a pipette, siphon a teaspoon of liquid from the bottom.
5. Use the appropriate enzymes to separate bacterial and viral proteins. Re-centrifuge and isolate the viral DNA.

The resulting cloudy white liquid, I pipetted into a tiny plastic Eppendorf tube that held about half a teaspoon. Starting from a liter of bacterial culture, I had sequentially purified my sample down to 20 drops of a pearlescent white DNA solution.

I'd spent 20 hours on those 20 precious drops.

Exhausted and looking forward to completing the extraction, saving my work, and heading home, I was on the last steps before overnight refrigeration. As I held the open Eppendorf tube in my left hand and the pipette in my right, I suddenly wondered what time it was. Was I going to miss the last Orange Line train going home?

I flipped my arm over to look at my watch.

I wear my watch facing outward on my left wrist. Some people wear their watches facing the inside, so that only a slight bend of the elbow and a subtle shift of the wrist is enough to glimpse the watch face. In hindsight, wearing a watch this way is a terrific idea for laboratory technicians whose hands are often carrying delicate experiments. Alas, I did not wear my watch that way. To check the time, I had to turn my wrist completely over, so that my thumb faced the floor. And if I should happen to be holding something—for example, an Eppendorf tube containing the precious results of an entire weekend of work—that would also flip over.

I watched as the liquid ran out. Onto the floor.

I looked at the upside-down Eppendorf in my hand. I looked down at the floor and the fresh splatter of my hard work. I froze, as if by freezing, I could gently coax time backward for a second or two.

And then, from down deep...

"Gnnnnnn... NooooOOOO!"

My late-night-fogged brain registered *shock*, *denial*, *anger*, and *bargaining* all at once. As the lab was deserted, I felt safe to express myself: "F$*&@! S!#%! D&**$%!" I shouted eloquently.

Desperate, I dropped down and started using the pipette to suck up dirty droplets of DNA from the floor. In a few minutes I had recovered about a third of the liquid, now brown-tinged, back in the tube. Resigned, I put the tube in the fridge.

No time to fret, no time to start over. Nothing else to do. I shrugged on my faded down jacket and hunched into the uncaring Boston winter, my dreams fading. I clomped my heavy boots toward the Orange Line stop.

All alone aboard that last train, I hung my head and sobbed quietly. There was no way the sample would be any good. My friends were out getting wasted. My weekend was an entirely different kind of wasted. What was I doing here?

And I realized—I really didn't want to be there. I realized I was able to do the scientific work, but unlike some colleagues, whose creativity blossomed in the long hours in the lab, I was not very good at it and missed being around people.

Today, thinking back and considering the book *Stumbling on Happi-*

ness,[1] I remember how author Daniel Gilbert says that a career that makes you happy can be assessed in three questions:

1. Is it meaningful?
2. Is it pleasurable?
3. Are you good at it?

I remember this as MPG: "Meaningful, Pleasurable, Good-at-it."

Scientific research was really meaningful to me, but I was neither enjoying the work nor was I particularly good at it. But abandoning the idea of scientific research put me in a tailspin in my junior year of college. *What would I pursue instead*?

The idea of medical school had always been my parents' dream, and not my own. Naturally, I had resisted, both because it was not my idea and because I thought basic science research was a higher calling.

On the other hand, as a sickly kid, I enjoyed visiting my pediatrician: a kind, wise and trusted ally. I also recognized that a doctor had to master both the science of medicine and connecting with humans. After examining my own feelings, and looking past the parental pressure, I realized that I did harbor a desire to be a doctor. I knew it would be a long and difficult road. In the following weeks, I embraced the decision and made plans to apply to medical school.

Hindsight is 20:20. At the time, my DNA crisis was a disaster. Looking back on my choice to pursue medicine, and subsequently medical informatics, I could not have chosen a happier path: one that let me care for patients and face scientific, technological, and psycho-socio-political challenges at the edge of my skills. My work is meaningful, pleasurable—and I dare say—I'm good at it.

The Lesson Learned

We all come to medical informatics from different directions. Some of us were programmers who later went to medical school. Others were doctors looking for a systems approach. Still others come from quality improvement and find that computers are a perfect tool for their work. Personally, I thrive at this intersection. Is this true for you as well?

Fall Down, Get Muddy, Do Informatics

When you first come to informatics, as a physician or APP, you may be lucky to join an informatics team and work on a project that is successful. More likely, though, you will be alone, trying to "do informatics" while striving to fit in with a colony of non-clinical IT folks or a pride of non-technical clinicians. You are likely to suffer setbacks, pushback, and resistance to change.

Perhaps you've heard some of the same gripes I have heard over the years:

From **surgery**: "Why do we have to initiate the surgery case request order in Epic? We never touch the EHR. We have people for that. This is a waste of time."

From **primary care docs**: "We are drowning in patient messages. I spend my evenings answering questions after an exhausting day. Can't you fix this?"

At the **medical directors** meeting: "Nope, we are *not* sharing progress notes with patients. They can't understand medical terminology. That is a terrible idea."

How can one keep going in the face of such discouraging words?

One day, coming home after a particularly challenging day at work, I unburdened my sorrows to my spouse. She told me, "Your job is always the same: Fall down, get muddy, eat garbage. Then you apologize and go figure something out. The next day, you get up and go do it again. Is that... fun?"

This stopped me in my tracks. Every part of her comment is true.

Thousands of doctors and nurses and millions of patients use the EHR every day. Informaticists are constantly building or adjusting the EHR tools. Occasionally, something you built lands like a dud. Everyone hates it.

For example, in 2013 we had the clever idea to create a list of insurance-approved reasons to obtain a radiology test. This was in response to a proposed national regulation called Radiology Indications of Use. A reason would accompany every order, and a click, we reasoned, was better than a few seconds of typing. You'd just select whichever common reason for getting a head CT scan. Is it for "severe headache?" Or "motor vehicle accident"? Or perhaps a "concussion"? Whatever you chose, it would already be insurance-approved!

Unfortunately, the list of approved reasons stretched to 80 lines. There are a *lot* of possible reasons someone might need a scan, and the "easy list" became a multi-page scroll. How could you find what you wanted? Would you look for "headache, severe" or "migraine," or (borrowing the patient's description) "worst headache of my life"?

That week, our organization had a sudden outbreak of "Acromegaly." This very rare medical condition unexpectedly became the most common reason for ordering a head CT. Go figure: It was one of the first reasons listed when sorting alphabetically.

We removed the new "indications of use" lists and acknowledged our design failure. "Sorry," we said, "our design was ineffective and the feedback was immediate. We will take this back and redesign it." We fell down.

When a problem with the EHR arises, we go to the bedside to investigate. We observe and interview to deeply understand the clinical situation. We study the possible technical solutions. We dig into the soil. We get muddy.

We are recipients of a lot of frustration... the "garbage" we eat are the criticisms of the EHR and associated workflows. It is useless to complain "But it's not my fault!" It doesn't matter. If you can avoid feeling attacked, you can think clearly. Some might say, eating garbage is natural to a PIG. Ok, that's a stretch.

And yet, we still have to do informatics. Informaticists listen to the frustrations, gather our clinical, technical, and emotional skills, and apply them to solve problems.

One might think that informatics is powered by technical expertise. This is not true. If you think informatics is purely about technical skill, come see me in a few years, after you have put together your own failure resume (see **My Failure Resume** on page 242).

Instead, informatics is powered by psycho-socio-political skills.

Communication, relationship building, and change management skills are the foundations of informatics.

Furthermore, to have a successful career, an informaticist needs good habits. Later chapters in this section will show you some of these good habits.

Informaticists also need their own support system, such as friends and colleagues who believe in a well-designed EHR and a thoughtful informatics team. You must be able to commiserate about your hard days, gather strength and insight from others also doing the hard work, to be back at it the next day.

The Lesson Learned

Informatics is messy and hard. You must have good habits and support for yourself to avoid an informaticist's burnout. All informaticists fall down, get muddy and eat garbage. Great informaticists turn muddy fields into rows of data, efficient workflows, and clear insights. We don't always succeed, but when we do, it is a thrill. *This* is Doing Informatics.

My Judge Shoots the Second Arrow

My parents brought me to the US when I was just a year old. I grew up in New York City, then Tallahassee, Florida. School wasn't challenging; I would often finish my homework before the end of the school day. This did not satisfy my mom, who would often read my textbook and make *more* homework for me. This was true injustice to a young boy whose friends were playing outside within earshot.

In sixth grade, I brought home my first report card. It showed five As and one B. When my mom saw it, she got quiet. "Show this to your dad when he gets home."

I thought nothing of it. My mom's words seemed odd, but I had not yet felt the weight of parental expectations. And then my dad got home, looked over the report card, and asked me to come to the dining table. We sat across from each other, and he spoke quietly. "We came to this country so you could have a better life. In America, there is more opportunity. We wanted this for you, our son. There are no other Chinese here in Tallahassee. We represent Chinese people to these Americans. The only thing that I give you, that is important to me, is my last name: Lin. Did you know that the Lin family history goes back 4,000 years? Here, the only thing Americans know about our family, about Lins, and about Chinese people, is what you do. The only thing I ask is that you do your best to represent my name. And you bring home a 'B'? This is your best?"

A heavy, quiet sigh. A slight shaking of head. He closed the report card, stood up, and left.

I was crushed. I would have preferred a spanking or some yelling. I had disappointed my father, and not only that, but... the entire Chinese civilization?!

I was 13. That was the last B I ever brought home.

My father's statement was, perhaps, a little unfair. But it certainly produced results. (This story is also an unfair caricature of my father, who was a great dad in so many ways.)

In conversations with other first-generation immigrants, I heard that versions of this same lecture have occurred in many households, spurring both exceptional performance... and misery.

Having now become a father myself, I have tried hard to avoid instilling this same fear and loathing in my own children. As it turns out, there is more than one way to raise a solid, hardworking kid. (Though it can help to tell a story about how mean their granddad was...)

In subsequent years, I have studied the mindfulness literature, and one concept jumped off the page at me: the internal voice of The Judge. My Judge sits on my left shoulder and says, "So, you got a B? This is really your best work?" This quiet voice has been motivating me my entire life.

Yes, guilt and self-critique are powerful motivators. Are you powered by a similar engine? Do you have a Judge inside your head that constantly disapproves of your procrastination, inefficiency, or less-than-perfect work?

This self-critique is like touching a hot stove. "Ouch!" The hand reflexively draws back. The guilt this triggers, taken in a small dose, can lead to impressive results.

On the other hand, if the internal criticism is loud and ceaseless, it is like leaving your hand sizzling on the stove. Your whole being simmers in guilt, anxiety, and lasting unhappiness.

What to do? Recognizing the existence of the internal Judge is a big step. Instead of telling yourself, "I'm not any good at this," you can instead assign that voice to The Judge: "My Judge says that I am not any good at this." What a big difference in these two statements. It's like having a volume knob for self-critique.

I have gradually learned to find this knob and turn the volume way down. "Yes, my Judge says this, but my Sage has something else to say." Just like you have an internal Judge, you also have an internal Sage, who sounds a lot like a wise mentor or colleague you've known. If you have an internal system of self-critique, shouldn't you have one for guidance and encouragement?

Shooting the Second Arrow

The Superman or Superwoman standard among medical professionals is alive and well. And it is killing us. Medical professionals show higher-than-average rates of depression, anxiety, and suicide. We are all simmering in our imperfections.

One evening, I sat at the kitchen table after dinner, obsessing over an event from work. A meeting had not gone well, and I was thinking I could have handled it better. I mentioned some of my frustration to my son.

Son: Don't shoot the second arrow, Dad.

Dad: What was that?

Son: Don't shoot the second arrow.

Dad: What does that even mean?

Son: Have you heard of *The Book of Joy*?[1] It's a conversation between Archbishop Desmond Tutu and the Dalai Lama. I recently finished reading it, and they talked about the "Second Arrow." The First Arrow is the original injury, the thing that hurt us in some way. Then the Second Arrow is the one we shoot at ourselves, by obsessing over that event, rubbing it in, and telling ourselves that we are not enough. Don't shoot the second arrow, Dad.

I had no response to this. Indeed, I had never heard of the Second Arrow, which, as I later learned, is commonly attributed to Buddhist philosophy. I spent the next couple weeks reading *The Book of Joy*. And I concluded that, yes, I was shooting the Second Arrow.

First, I was offended that some high school punk (my clever son) was giving me advice. Then, I spent some time acknowledging that one's own child can offer helpful advice, and I could feel the tension drain from me.

Finally, I was grateful, and maybe a little proud. Since then, I have told colleagues and my own mentees what my son taught me: "Don't shoot the second arrow, duh."

> **The Lesson Learned**
>
> If, like me, you grew up scrutinizing every lapse in judgment or failure to exceed expectations, then you're likely walking around

with an arrow or two stuck in you right now. Most likely, it was your Judge who strung the bow. Our caveman brains survived by obsessing over threats. But in the modern age, such chronic stress is destructive. If you have an internal Judge shooting at your stumbles, balance them with a Sage who applies salve and supports you.

My Failure Resume

Even before the COVID pandemic, we struggled with the self-sacrificing culture of healthcare. Unlike other businesses focused on profit and self-interest, selflessness is the core of medicine.

Let's face it—we healthcare professionals do a poor job of caring for ourselves. We are constantly engaged in reducing suffering, treating illness, and saving the lives of patients. We work endless hours and resist the idea of taking a vacation in fear of being labeled "selfish." Because the stakes are so high, the pressures rise to match. If we are not at our best, offering less than total commitment, we could be considered a failure.

"Failure" is a weighty word in our world. It can represent lives lost. For many of us, the idea of "failing" is terrifying. But is that reasonable? Failure is unavoidable.

In addition to my clever son, who introduced me to the idea of the Second Arrow (see previous chapter), I also have a clever daughter. We were talking about the pressures we put on ourselves, and she introduced me to the "failure resume."

My daughter had come across this idea in her reading on mindfulness. If a typical resume is a collection of one's successes and achievements, a failure resume presents one's biggest defeats. It is a celebration of setbacks. And it reminds us that everyone, including successful people and senior colleagues, have all experienced many failures. I loved this idea.

In short order, I wrote my own failure resume, which I posted on my blog. (My failure resume is included at the end of this chapter.) I thought it would be a funny story to tell and that my friends and I might get a quick chuckle out of it. Little did I know.

In short order, my failure resume post became the most popular entry of my blogging career. I cannot say for certain what made it so popular, but I think it served as a counterpoint to the dangers of Social Media Syndrome.

Social Media Syndrome is the persistent pressure to appear invulnerable and perfect. We scroll through our friends' posts and see them scuba diving in Fiji, or skiing in the Alps, or hiking to the Everest base camp. Everyone else's life is better and more impressive than ours.

Naturally, no one posts about the divorces, the layoffs, the investment losses. To be more accurate, everyone else's highly curated online life is better than mine.

What better way to combat this than purposely sharing a resume of your failures?

In the years since I posted my failure resume, many colleagues have told me they wrote their own. It is a powerful exercise in vulnerability, and I have enjoyed building friendships with those willing to admit their failures. We all have failures. No one is excluded from this club.

The Lesson Learned

When a respected leader writes a failure resume, everyone benefits. Leaders who write one seem more approachable. Our failures trigger empathy, and empathy is the root of human connection. Our workplaces can be less unfriendly and more joyful.

CT Lin MD, FACP, FAMIA

Chief Medical Information Officer, UCHealth
Professor, University of Colorado SOM
8/2020

Profile

CT has failed at so many things it isn't funny. Or maybe it is.

Make old mistakes less often than new ones.

Sometimes you have to outlast your naysayers.

Non-Skills

Handles confrontations poorly.

Midnight emails set a bad example for others.

Binge-TV, -gaming, -posting are maladaptive skills.

Inbox overflowing, and not from "love."

Contact

LinkedIn	*linkedin.com/in/ctlin*
Twitter	*@ctlin1*
Publications	*researchgate.net/profile/CT_Lin*
YouTube	*youtube.com/ctlin80045*

Ukulele

Plays ukulele because after several months, unable to figure out guitar. See the bad singing on his YouTube channel.

My Failure Resume

1976 -- 2000

1976: Got a **B** in middle school. Had a quietly forceful lecture from Dad: "We came to this country so that YOU could have a better life. The only thing I give you worth anything is my last name. In America, YOU represent over 4000 years of our history. So, you got a "B". Is this how you represent your family, and my name? I was 13.

1985: Applied to 12 medical schools, was **rejected** by first 11.

1989: Residency: Applied to 14 residencies for Internal Medicine, was selected by my **8th choice**. After bemoaning my fate, ended up meeting my fiancée there.

1992: When fiancée was accepted to pulmonary/critical care fellowship in Colorado, I quit my job, got married, moved, had to find a new job. Interviewed at all the major health systems in Denver, no one interested. Only found a job at UCH **by accident**.

1997: **Chief Complainer** about terrible state of computers at Univ. Colorado, sent provocative email to administration. Reply: "We will pay for 4 hours of your time per week to explain to the docs about computer systems." I took job as IT Liaison.

1998: **"Too junior"** to be promoted to Medical Director of IT, passed over in favor of another guy. 1999: other guy quit, no other candidates, I got the job.

1999: One RN and I were training docs to use the EHR (3M Clinical Workstation). I was **politely asked to go away**: they were busy being doctors. Hospital leaders' response? "Go find other clinicians who might be friendlier to you."

2000 – Present

2001: Conducted a research study (SPPARO) on sharing test results and doctors notes with patients online. Despite positive results, our clinical leaders **chose NOT** to adopt this. Only years later did I understand that changing a large organization is more about persuasion and relationships than it is about p-values and science. It was **16 years** after my original research before we deployed Open Test results and Open Notes.

2002: Seeking promotion to Associate Professor, the committee **declined** my promotion 4 times before conceding that I deserved the title.

2003: Switching from 3M to Allscripts EHR. Only have $400k instead of 2.5M to install an EHR. Vendor: 'We don't recommend it, but ok.' Result: 17-year deployment plan. **Ended up with worst case**: a stalled project with 40% of clinics on EHR, 60% on paper. With dual record systems, everyone needs to pull BOTH charts.

2007: Introducing APSO notes (inverted SOAP notes) for improved readability of EHR notes. I thought it was a great idea. Convinced 80 interested colleagues to try it. APSO use rate during "opt-in" phase? **16% (almost no one).** Years later, I was able to gain clinical leadership acceptance and make it near-mandatory: "You want to use our new Epic EHR? It only comes with APSO notes. Sorry." APSO use in Epic? 90%.

2012: Radiology Appropriate Use Criteria. Created a list of "appropriate indications" inserted into radiology orders. Unfortunately, it was a long scrolling list alphabetically sorted, hard to read. Soon after, there was a surprising outbreak of patients with "indication: ACROMEGALY" as it was the first checkbox on the list. **Shut off this project** after 1 week. (Update: successful go live 7/2020, 8 years later!)

2017: Eastern European Hackers attacked and disabled national Transcription Service computers. Hundreds of surgeons lost a week of dictated Operative Notes. It was 3 weeks before transcription service restored. **Vicious Rumor**: "I'm pretty sure CT Lin did this, to force us all to type in his f-@#&-ing EHR." Somehow kept my job.

2019: After 3 years of weekly blog posts on leadership and informatics, my most popular post? THIS FAILURE RESUME.

The Social Media Power Law

Humans are complicated. Put several humans together in a meeting, and the social dynamics can be hard to predict.

As a leader, it is difficult to gauge the success of a meeting you led. You're focused on managing the agenda, ensuring a clear discussion, and then establishing a clear decision. You cannot also be focused on all the other moving parts—whether all members are participating, whether they find the meeting to be valuable.

I was formerly the president of the Allscripts TouchWorks Academic User Group for the electronic health record that our organization was using in 2005. (I tried to name our group "Group of Users—Academics, the National Organization" (GUANO). Shockingly, my rogue naming proposal was not successful. But that is another story.)

There are many challenges to deploying a complex electronic health record. It was natural that user-groups would form to share good ideas. Our academic health system joined 17 others in forming one such group. We met monthly by telephone conference, with one annual in-person meeting at the TouchWorks National User Group.

One year, it was my turn to preside over the group, assisted by a vice president and secretary who helped manage our group meetings. On our monthly calls, we discussed best practices, shared our victories and defeats, and agreed that it was tough to get busy doctors to use the computer to care for patients.

Around 25 people dialed into our monthly calls. We three officers would discuss topics of common concern: how best to encourage brevity in progress notes (and decrease so-called "Note Bloat"), how to design more efficient orders for prescriptions and radiology tests, how to most efficiently search the chart, how to encourage our EHR vendor to be more responsive to our requests for software improvements. And so on.

As president, I led many of these discussions, but I failed to get other members to contribute to the conversation with any regularity. In a typical hour-long call, the VP and secretary would chat with me, and one or two other members would occasionally chime in. The remaining callers would remain silent for the entire hour. Despite my best efforts to engage them in dialogue, I would often be met with deafening silence. I thought to myself: if only five of our 25 members are speaking, that means 20 people are not getting much from our discussion.

At the end of the year, we met in person at the TouchWorks-hosted national user group meeting, and I apologized to the 25 or so members in attendance. "Sorry we didn't have conversations with more participation." Surprisingly, almost everyone spoke up to say that our monthly meetings were very useful!

"Why would we continue to dial in and listen every month, if they were not interesting and useful?" They remarked that our discussions always centered on topics of common interest to them. I was briefly floored. But in hindsight, taking into account the social media power law, this now makes sense for telephone conferences.

The concept is as follows: Only about 1% of users on social media actively create content. Another 9% will comment, edit, or contribute occasionally, and the remaining 90% will be lurkers (those who consume the content without adding anything). In other words, only a few individuals of a group will be active creators, and a small number will contribute, but most will be silent. The important point is: Regardless of participation level, *everyone* benefits from the conversation. This came as a relief, and it explains a lot about the success of online communities.

How might we apply this to in-person and online meetings?

In the book *Here Comes Everybody*,[1] we learn that Wikipedia, with millions of online entries, follows this power law: Only 1% of the community is content creators, and another 9% leave thumbs up or comment on others' writing, yet everyone, including the silent 90%, gain value from its use. Without these stats, one might wrongly presume that Wikipedia had a much broader base of contributors.

In meetings, there is often a similar 90:9:1 ratio of participants. When you exceed this and more participants are engaged, count yourself fortunate. You are doing something right.

With in-person meetings, we can see body language and hear tone of voice. In my experience, this facilitates human connection and allows for better discussions. Online-only meetings require more forethought, and participation may drop below this ratio. Here are some lessons I've learned to raise participation in online meetings.

Having video on during meetings keeps everyone's focus. But so will a clear, focused agenda with topics led by different members. It's also helpful to have someone different from the leader assigned to watch participation by members and encourage more people to speak.

An interesting feature of online meetings is the added presence of the sidebar chat, thumbs up, and emojis. If the chat is managed effectively, many more participants can write in or communicate their ideas while others are speaking. In the case of an occasional long-winded verbal participant, others can collaborate on a parallel conversation; in one case, I saw members coming up with guiding principles while the current speaker droned on, defining the problem.

Another tool is the breakout room. When meeting in person, it takes coordination to split people into small-group discussions and then reconvene. This is much easier when meeting digitally. Assigning three to four people to a single task or topic and splitting them off will let them make rapid progress on their issue and bring good ideas back to the larger group.

A hybrid meeting, with only some members attending online, is the worst of both worlds. Sidebars in the chat can be missed by in-person participants, and conversely, online participants have difficulty "chiming in" to the in-room conversation. I avoid this situation whenever possible. Instead, I shoot for periodic in-person meetings to strengthen the in-person rapport (twice a year in for my Large PIGs, with other meetings online-only).

Regardless of how we are meeting, I use the social media power law as a benchmark. I watch the participation and ask myself, how are we doing? Have we heard from at least 9% of the group? What might we need to adjust to increase effective internal collaboration?

The Lesson Learned

90:9:1 is not an excuse but a forgiveness. It is also a benchmark to exceed for your online and in-person discussions. Consider specific tactics to increase participation. The more brains involved, the better your decisions.

Why You Should Write a Gratitude Letter

"I'm calling to tell you I'm entering hospice. I suppose this is goodbye."

It wasn't a call one gets every day.

The caller was Dr. Fred Platt, a beloved mentor and colleague. He authored many peer-reviewed articles and books on physician–patient communication (see **The Anxiety Problem with Test Results** on page 188). Several decades ago, he had helped me get started in academic medicine.

After this call, I decided to go visit him in person, but I did not know what to say.

I found that expressing my gratitude was as simple as writing him a letter. In the wellness literature, a gratitude letter is a well-established intervention. Here's how they work: Write a letter, tell someone how they made a difference in your life, and thank them. It's that simple. Go see them in person. Read the letter out loud to them. The research tells us that both the giver *and* recipient will have their moods boosted, sometimes for months.

Best of all, giving gratitude is free.

I wrote Fred a letter. It did not take long. It was easy to recall the moments when his gentle hand helped form my identity and shaped my career. Then I made an appointment and drove to see him. He only had the energy for a 30-minute visit, but he loved the letter enough to have me read it to him twice. We both cried. He'd had a tough few years, and we had lost touch. I was so happy to provide him with some joy and meaning in his last days. I have thought about him often in the years since, and so grateful I had a chance to tell him what he meant to me.

Here is the letter.

Dear Fred,

Thank you for the chance to tell you how you've changed my life.

I met you at the Bayer Institute CPC Workshop: Clinician-Patient

Communication to Improve Health Outcomes. You facilitated a group of 16 junior faculty in Internal Medicine in 1997, and taught me, among other things, reflective listening:

"So, it sounds like you're having some belly pain, and it is going down to the right side, and you think it is... gout? Do I have that right?"

I learned about reflecting not just data, but also ideas, feelings, and values, and it changed my life.

I learned that I could reflect IDEAS: "It sounds like you think this is gout?"

I could reflect FEELINGS: "It sounds like you are really frustrated with me!"

I could reflect VALUES: "Sounds like what is important, is that you can't miss any more work."

It rocked my world.

I learned these skills for myself. Then, I learned to teach these skills and follow your footsteps. I did not know it at the time, but these skills changed my career path.

You took me under your wing and gently taught me tools for difficult conversations with patients as well as colleagues: "On the one hand, you really hate the electronic health record... On the other hand, these tools are mandatory, and I can show you some tricks... Maybe we could work together on this?"

You cofounded Foundations of Doctoring at University of Colorado. We first tried teaching communication skills in lecture halls. 160 students came to the first lecture. How many came to the second lecture?

Five.

A lecture hall is a terrible place to learn interpersonal communication.

We redesigned the course until it really sang: one standardized patient, four student learners coaching each other, and a facilitator guiding Invite, Listen, Summarize: I.L.S.

Students told us, "This is the most important thing I've learned in medical school!" I use and teach these skills today with incoming students.

You generously asked me to coauthor communications papers, including the series "Words That Matter" in the Annals of Internal Medicine. We discussed many delightful cases, including the case of my patient who worked in a bakery, ate too many pies at work, developed diabetic coma and crashed his car, and wanted to claim workman's compensation, prompting an angry outburst from my resident. "No! You can't do that!"

In my field of informatics and electronic health records, you taught me that it wasn't about the technology but instead about building strong relationships with excellent communication skills.

I've learned over the years that I have an internal Judge voice, who sounds suspiciously like my father. I also have an internal Sage voice, whom I call "Fred."

Thank you, Fred.

With Love,
CT Lin MD, Professor of Medicine, CU School of Medicine

The Lesson Learned

Don't wait for a call from hospice. Gratitude is free. Whether making a brief, sincere comment, or writing a full gratitude letter, building a habit of expressing gratitude is an overlooked superpower.

Every one of us has benefited from the generosity of others. I challenge you, now, to think of one person to whom you are grateful. Who is that person? What might you say to them? Write that letter today.

Medicine Is a Performance Art

There is a faded black-and-white photo of me at about age 10, standing with my cousins somewhere in Taiwan. We were visiting the island where I was born to see my extended family. The Lin family in Taiwan is a big one.

In this photo, you see my plaid short-sleeved shirt, my patterned tie, my shorts, and my white socks pulled up tightly, as high as they'll go. If you look closely, you can spot two mysterious lumps—one in each of my socks. What *are* those?

I had stuck my prized possessions, Matchbox cars, into my socks for safekeeping. I wanted to bring my cars with me when we went out.

My mom would volunteer to put my cars in her purse as we left. NO WAY. I needed instant access. I could hold them in my hands, but I might get distracted and let go and lose them. I could put them in the pockets of my shorts, but they were too small, and I worried they would fall out. Where could I safely store them...?

My *socks*. The elastic would keep them safe, and they would be out of the way but still instantly available for emergency boredom situations. It was perfect.

We often visited relatives, and having been raised in America, I only spoke a few words of Taiwanese. Worse, there was never anyone near my age—all my cousins were much older—or we would be out sightseeing, where all the signs were in Chinese.

In such moments, it was time to get out the trusty Matchbox cars. Any time I was bored, *zip*, out came the cars to race on the banister, across the checkered floors, on the edges of tables (the sheer cliffs of Speed Racer's rally race), me revving the engines under my breath (vrmmmm! Rrrrr-rmmmrrr!) and screeching around corners (rrrRRRRRrrr!) What more could a boy need?

In hindsight, my car-filled socks demonstrate two of my core qualities: First, I am pragmatic. I see nothing wrong with getting the full storage potential from my footwear. Second, even then, I was happy to ignore the group and lose myself in a tiny world of my own.

In other words, I am an introvert.

My introversion served me well when it came to the grueling studies of medical school, but the problem is—surprise!—medicine is a performance art. Being shy and withdrawn is a liability. This took me some time to realize.

I began my third year of medical school incredibly excited to put two years of book learning to use. I had passed all the examinations testing my knowledge of the dozen medical texts on my bookshelf.

My first clinical rotation was surgery. In preparation, I had splurged and bought black Reebok sneakers. I had heard we would be on our feet for long hours, walking many miles inside the hospital. I also bought a pair of Ray-Ban sunglasses. I was leaving behind my books and becoming a clinical student. I was going to be cool.

At the end of the two-month rotation: I had worn out the Reeboks, and I never once wore my sunglasses. We were up before dawn and home long after dark.

The surgery rotation itself? It was an absolute surprise and joy. I did not expect to enjoy surgery; my hand-eye coordination, although adequate, was not comparable to that of my classmates, who had instantly mastered tying precise surgical knots with hemostat clamps. I was more interested in complex illness than in cutting. And yet: my entry into the operating theater was magical.

"Step over here, change out of your clinical clothes and into some scrubs, mask, cap." I did so and stepped into the foyer adjacent to the operating room. I was in sensory overload: the smell of disinfectant, the temperature drop, the hum of the high-volume air exchanger, the bright lights, the surgeons in the corner bantering and scrubbing.

"Okay, move it, get over and scrub your hands down to your elbows, using this scrub, exactly as I do it." Eyes wide, nodding mutely, I mimicked every motion of my resident. "Hold your hands upward. Make sure the water drips toward your elbows, not back onto your hands."

"Keep your arms like this, back your way into the operating theater." I was met by a nurse, who held out a gown, arms slid through, gloves snapped in place, gown efficiently tied in back.

"Step up here, and do *everything* that your resident or attending tells you."

The next two hours were a mixture of terror and amazement, watching the smoothly choreographed dance of flashing instruments, blood, suction, layers of tissue revealed, the offending tumor removed or defect repaired, questions posed to me about anatomy, which I stumbled through, and a vow that I would read everything about the operation I had just seen once I got home. (Despite that vow, I got home, gobbled a quick TV dinner, and instantly fell asleep—a pattern that would repeat for the next two months.)

In subsequent days, I woke up at 4:00 a.m., got to the hospital to see my post-operative patients at 5:00 a.m., gathered relevant lab results and reviewed the records of the last 24 hours, presented the cases to my team at 6:00 a.m., discussed some pathophysiology over a quick breakfast, and headed to the OR at 7:00 for another breathtaking operation. My team was amazing! Eight weeks flew by. I had learned to see patients in clinic, remove sutures and check for post-op wound healing, interview patients in the hospital preparing for surgery, scribble daily progress notes, improve my examination and presentation skills, read voraciously, and generally feel exhilarated by each passing day. I began to surprise myself, thinking that I might like a career as a surgeon.

It was a crushing blow, then, to receive my grade from the attending surgeon: "Below expectations—Pass."

I read my evaluation. "Obviously not interested in surgery. Does not speak up on rounds. Knows his patients well when presenting on rounds, but does not ask questions about other patients, does not participate in general discussion on anatomy or pathophysiology."

I was devastated. Having aced high school, college, and the first two years of medical school, here I was with a near-failing grade. I thought I had done well in the rotation. I showed up and always completed my pre-rounds. I paid attention and was always prepared to discuss everything about my patients; I brought research papers about my patients' particular versions of diseases; I held retractors and answered "pimping questions" in the OR ("Mr. Lin, is this an artery or a vein I'm pointing to?" "Um, an artery?" "Wrong, it is a nerve."); I did errands for team members; I stayed overnight when on-call; I only went home when the work on my team was done for the night. And yet, in that moment, I learned the grading criteria for clinical rotations.

In a moment of clarity, I realized that being a physician was a performance art and not a bookworm task. And if being extroverted is what it would take to be successful in medicine, I was going to do it. I fashioned an Extrovert Cape for myself and put it on every morning in the pre-dawn before rounds. I would ask questions about other students' patients. I would approach my attending on rotations and ask, "What will it take for me to get an outstanding grade on this rotation? Because I am very interested in this specialty, and I'm willing to do the work." Once home, I would collapse into my natural state: sitting, reading, reflecting quietly, and a long sleep to recharge.

Over the years, this adaptation served me well. I would naturally learn by reading and introspection, but as soon as I arrived at the hospital, I appeared curious, asked questions, was extra helpful to my intern, my resident, and my attending. I would volunteer for thankless tasks. Ultimately, my assessment proved my suspicion: medicine *is* a performance art.

Even now, some colleagues are surprised that I consider myself an introvert, as my Extrovert Cape is on all day, every day. I have perfected the performance.

On the other hand, my children understand this about me. "Oh, did you have to lead a bunch of meetings today? It's Introvert's Nightmare Tuesday, again, Dad!"

My ukulele also had its origins in this introvert/extrovert performance dynamic. Despite my inner introvert's strong objections, my extrovert persona pushed me to bring my uke to work and play my EHR parody song for the first time. (More on this in the next chapter.)

Behaving like an introvert is a detriment in medical school and in the practice of medicine. Introverts must make an effort to ask questions, verbally express interest, and offer to help. This extends to the practice of medical informatics. Extroverts, who interact more naturally, have advantages. Learn to adapt. It's not enough to know your stuff; you have to *show* your stuff.

Don't get me wrong: introverts bring great value to the practice of medicine and informatics. We work well alone, we persevere despite setbacks, and we think deeply and insightfully.

Recognize your own natural inclination toward introversion or extroversion. Learn to exercise both qualities in yourself. Medicine rewards both extroverted collaboration and quiet, introverted expertise.

The Lesson Learned

Whether you are introverted or extroverted, are there ways you can help your team by coaching each other? Try the opposite end of the spectrum from your natural behavior.

A Ukulele in Healthcare

"What is this tiny guitar doing here?" I wondered as I went through the closets of my rented Hawaii condo. That discovery was the start of a peaceful 10 days on the Kona Coast. After several evenings learning chords, strumming, and an "open to the public" ukulele lesson at the local shopping mall, I was hooked. I had to bring home a ukulele. Little did I know how that "tiny guitar" would change my life.

Back home, my ukulele became an instant end-of-day release. Strumming and learning songs became a routine joy. Whenever I glanced at my uke, sitting in my living room on its stand, I could not resist picking it up for a few strums. Just a few strums, and already the day is better.

I began looking up ukulele lessons online and found Ukulele Underground,[1] with hundreds of free and paid lessons. I also joined the Denver Uke Community (DUC). It turns out that many US cities have ukulele communities with regular meetings. The DUC provided beginner song books, and my children and I joined every month for a couple years. I had no aspirations for ukulele stardom. I was happy participating in that small, tight-knit community and strumming at home. I never thought I would play in public, or at work.

That winter, we were driving home from a ski trip with the kids asleep in the back, and Elton John's "Rocket Man" came on the radio. My wife turned to me and said, "'She checked my bag last night, preflight.' That's you! Aren't you running *preflight* checklist sessions the night before Epic go-live with your docs? You have to write this song! Come on! I'll even give you the name of the song: Epic Man!"

"No," I said. "I don't write parody songs. No way." But the song stuck in my head, and I ended up writing the lyrics at the kitchen table that night. I played it at home for fun for a few weeks. Then, I put the song away and forgot about it.

A few months later, in February of 2011, it was go-live day for the Epic electronic health record for our 600 doctors and 2,000 nurses. Our team of analysts and trainers had spent 18 months getting ready.

The go-live Saturday was a beast: I walked 27,000 steps that day, encouraging docs to use the computer and apologizing that their pens, prescription pads, and paper charts were now gone. Often, I would kneel next to a colleague to help out as they struggled to admit a patient, to prescribe home IV antibiotics, or to write a complex radiology order. My aching knees told me that I should have done more deep-knee-bends to prepare.

Like many organizations deploying an electronic health record, we set up a "command center" staffed with about 50 red-shirt-wearing IT trainers and analysts who were EHR experts. Our "red-shirt" IT support staff answered the phone and fielded over 7,000 complaints from angry doctors and nurses in the first seven days.

"The printer isn't working."

"Where do I find the f***ing orders?"

"What's my password again? Why do I even *need* a password?"

By that next Friday afternoon, the team was completely exhausted. The combined haze of body odor in the poorly ventilated command center was... ripe. They needed a pick-me-up, so our IT team leaders arranged for the CEO and CIO to visit and cheer them on. As CMIO (Chief Medical Information Officer), I wanted to offer my own support. Was there something I could do?

Being an introvert in a position of leadership, I was always looking for a new way to put myself out there and wear my Extrovert Cape. Was I dumb enough to take my ukulele to work and play "Epic Man"... for the 50-member command center? What if they laughed at me?

Ultimately, I turned that same thought around—their laughter is what I wanted. I wanted them to forget, even for a moment, the swirl of strong emotions around them. After all, an entire medical center and the daily lives of these physicians and nurses had been upended in the move from paper to digital. The force of that anger and fear from loss of routine was directed at my hardworking, never-complaining IT colleagues.

I almost didn't bring my ukulele that day. As I was leaving home, my teenaged daughter spoke up from the breakfast table: "Omigod. You're not taking that to work, are you? You know that you're a terrible singer." I hesitated and almost put the instrument back, but then I took a look at the chasm in front of me... and jumped.

There is no recording of that day, but I ended up playing "Epic Man" on the "back forty" during karaoke afternoon at the national Epic meeting. Just search YouTube for "Epic Man CT Lin." Or, aim your smartphone camera at the QR code below.

The enthusiasm of the red-shirts was overwhelming, and their cheering was a celebration and release all at once. I saw more than a few folks smiling and shaking their heads.

It was deeply uncomfortable to put myself in the spotlight by sharing my fledgling skills at singing, playing, and songwriting. However, I think that my vulnerability let the team know it was okay to laugh, cry, and let go of the strong emotions they'd pent up that week. I'm glad I brought my ukulele to work that day. Since then, I've built a side-hustle playing ukulele parody songs about our growing EHR.

The Lesson Learned

To this day, I'm nervous when I stand up to play. The ukulele is a symbol of me stepping outside my comfort zone and injecting some vulnerability and fun at work. How might you do the same?

See and hear "EHR Epic Man" for yourself!

Grow a Strength or Patch a Weakness?

Here are some of my weaknesses: I'm terrible at writing grants and obtaining federal funding. Some of my colleagues are skilled at chasing federal funds, but I am reduced to asking local leaders for funding. When I fail to make a convincing argument, I find myself doing a tiny project for no budget (see **The Medium View: Accountability & Tuesdays to Write** on page 288), or I wait years to get a project off the ground.

I'm also terrible at the nuts and bolts of hiring and firing. I can encourage a struggling team member to improve, but I have difficulty removing a colleague who isn't a good fit (see **Grow Your Team: Square Peg, Round Hole** on page 166). I'm terrible at building budgets with capital and operational expenses offset by a formal return on investment. Thank goodness I've found finance and human resources colleagues who make up for it.

On the other hand, I believe I am strong at listening, being present for others, speaking and writing clearly, being a good advocate for patients, colleagues, and shepherding important longshot projects. Like everyone, my weaknesses are offset by strengths.

Yes, it is a cliché to suggest the importance of growing your skills. Yes, we should all constantly be improving ourselves. But that still leaves a difficult choice: Do you build further on your areas of strength, or should you seek ways to shore up your weaknesses? Do you maximize, or do you try to balance? Which should you pursue?

The short answer is, "both, duh." You already know this.

The longer answer: We can focus on our weaknesses to make improvements. For example, I have always been a strong introvert—given a choice, I will prefer to sit in a corner and read a book rather than join a party or a large discussion. But I also know that medicine, and informatics, is a performance art (see **Medicine Is a Performance Art** on page 252), and thus I force myself to be gregarious and outgoing. Extroversion is an important CMIO skill. Playing ukulele is an extension of patching my weakness. I still force myself to get out and give talks and play, and it has gotten a little easier over the years.

Furthermore, I prefer to stay introvert-quiet during meetings and only take out my uncomfortable Extrovert Cape and speak up at the end to do my "bucket summary" to help conclude a meeting successfully (see **"The More I Listen, the More My Patients Learn"** on page 216).

Even more important, do not ignore your strengths. Instead, take your strengths and make them extraordinary. In fact, I think this is easier and more rewarding. I can take something I'm already good at, such as writing a clear but overly long communique, and refine that skill and create a compelling one-page report (see **The Superpower That Is the One-Pager** on page 83). I love closing the door, laying out all the evidence, and crafting the perfect one-page summary. Practicing has made this better, and I enjoy it.

College admission deans write about being "spiky" versus "well-rounded." "Spiky" hints at the idea that someone with a deep interest or passion will develop an interesting expertise that can serve themselves or others, as opposed to someone well-rounded, who may dabble briefly in many areas. Your strengths have the potential to be the things you contribute to your field, the mark you leave on medicine.

You don't have to figure it out by yourself, either. Atul Gawande, in the book *Better*[1] and his corresponding TED talk, "Want to Get Great at Something? Get a Coach,"[2] points out that world-class athletes have coaches to help them win tournaments, so why don't doctors have coaches that guide them toward excellence in their careers? Why don't informaticists?

Even senior professors should have a coach or mentor and talk about ways to continue to get better. Thirty years into my career, I continue to meet with my own mentor. There is always someone who can offer you a new perspective or a piece of wisdom. Why stop learning and improving?

Books are another great way to polish your strengths and expand your knowledge. Sometimes I read to shore up weakness; often I read to learn how to further sharpen my best skills. I also use a book club for reading motivation and to get different viewpoints as I digest the new information.

Books are a powerful tool for improving oneself and one's team. (For a list of my favorite books, see the bibliography at the end.)

It is said that spending 30 minutes a day for 30 days is sufficient to learn a completely new skill or transform an existing skill.[3] How would that feel? Can you start today?

The Lesson Learned

Sometimes going from good to great makes more difference than going from bad to average. Don't ignore the joy in honing your strengths to be remarkable.

You Don't Want a Title; What You Want is Time

"I want to be an associate professor with an academic appointment."

"I deserve to be Associate CMIO."

"I am not recognized for my work."

Have you ever had thoughts like this?

Physicians are driven to achieve. Many of us were at the top of our classes in secondary school, undergraduate studies, medical school, then residency. That's at least 15 years of collecting certificates of achievement. The habit is hard to break. Even after our formal training is completed, the pursuit of excellence and desire to be recognized remains.

For those entering medical informatics as a career, it is natural to seek recognition or a title to distinguish yourself from a frontline clinician. Bad news: In the early years of your informatics path, titles are rare to nonexistent. But the truth is, you don't need one. Don't fall for this false flag. Titles are less important than your true first goal: protected time.

Work hard to arrange protected time, away from the treadmill of treating patients, to think about EHR design and the flow of information. In our organization, we have over 200 certified Epic SmartUsers and Physician Builders. Some of them do good work without protected time. However, the best predictor of someone who does informatics work successfully is someone who has protected time. Without protected time, the treadmill of being "clinically productive" and seeing more patients is unending. One is constantly exhausted, and it is difficult to find an hour or two to set aside and work on an informatics project.

First lesson: 20% protected time (or one day per week) is the minimum an informaticist needs to be effective. There are exceptions to the rule, of course, but this is my guidance to those just starting out.

Are you still coveting your colleague's title? Do you absolutely need something to put in your email signature? There are a lot of ways to get a title, if you absolutely must have one. Consider these situations.

Fake It 'til You Make It

At an east coast hospital in the 1970s, a couple of nephrology physicians began doing kidney failure calculations on computers. Believing that computing power would transform the practice of medicine, they dreamed up a new term and taped a scrawled sign outside their office: "The Department of Medical Informatics." This was a running joke for years—until the hospital's sign painters came through and made it official. Just like that, they were bona fide. From there, they started attending the faculty senate meeting to represent their newly christened department.

Wear it Proudly

My first job in IT was "Physician Liaison to the IT Department." I was told to "describe all the good things that IT is already doing for doctors. Use doctor-speak with them and get them to understand." It was a disappointing title. However, I worked at it, and I did well enough to be promoted to "Medical Director of Informatics." What changed? Not much. Titles mattered less than I had thought. As a "liaison," I had met with doctors and IT staff, listened, built trusted relationships, helped deploy a few EHR-related tools successfully, and showed people that I could find solutions to problems, or at least explain why something could not work that way. Becoming a Medical Director did not open new doors or increase my effectiveness. (My mom liked my new title, though.)

Make the Job Your Own

Even once I was Medical Director of Informatics, no one knew what an "informaticist" was, so I paved my own path. In addition to deploying electronic records, I constructed a research portfolio one project at a time. I worked with others to study the new idea of patients sending online messages to their doctors. In those years, when the EHR was underfunded and deployments were slow, we found the time to study EHR patient portals and publish results. Make the job your own.

The Lesson Learned

You may be hungry to distinguish yourself with an impressive new title. Set that aside and seek at least 20% protected time, the minimum an informaticist needs to be effective. When you are bestowed a title, use it to your advantage and don't worry about coloring inside the lines. Make your job your own.

On Self-Management

The THRIVE section began with a reminder to be kind to yourself, to turn down the volume of the Internal Judge who presides over every setback. We must accept that failure is inevitable and sows the seeds of opportunity.

At the same time, I know that healthcare professionals are driven to Get Sh*t Done. Our time is limited, but the demands on our time are not. Our clinical lives are highly structured by clinic schedules or inpatient census, one patient after the other. Clinical work is difficult, but clinical time management is easy: Finish evaluating and treating this patient, then go see the next one—they're waiting for you. Keep going until you are done.

On the other hand, informatics work is unstructured, sometimes to a disorienting degree. You sit down at your desk for four hours of protected informatics time. ...What do you do? How can you capitalize on that time and ensure it is productive?

This subsection is a collection of tools and approaches that will help you achieve more and grow faster in informatics.

Punctuated Equilibrium

It's not possible to go 110% all the time. And yet, it's easy to feel like anything less than 110% is insufficient. We are all afflicted with "Social Media Syndrome." It is very easy to hop online, look at the highlight reels of everyone else's lives—your friend vacationing in the Bahamas, your other friend publishing papers on AI, your friend being recognized as CIO of The Year—and conclude: my life sucks. I suck.

Despite this, I was recently asked, "How do you get so much done?"

To me, this question is a classic manifestation of jealousy, regret, and unfair personal self-critique. In these times, it is critical to recognize that any outward success has three hidden factors: sacrifice, innate advantages, and punctuated equilibrium.

Sacrifice. When struggling to complete difficult projects, I had to sacrifice. It simply isn't possible to do everything. Working hard to achieve something in your career means leaving something else undone. It might be a relationship, family obligations, hobbies, or personal goals.

Regularly pause and reflect. When you notice yourself being jealous of others who are "more successful," stop. Remember that they are sacrificing something, and likely something you are not. It may not be visible. You have made different choices, and that is okay.

Innate Advantages. Some of us have duties outside of work. Some of us have elderly parents or young children—or both—who require attention. It is likely that those who appear to achieve more also have fewer burdens. Forgive yourself and do not suffer the comparison.

Punctuated Equilibrium. My daughter's lacrosse coach once told the team: "You should not run at the same speed the entire game. There are times to jog around the field, finding position, saving energy. There are times to sprint all-out to get the ball, to get position on an opponent. Find the right time to sprint, and then go all-in."

In the 1970s, it was determined that biological evolution does not occur at a gradual rate. Instead, it describes a pattern of "punctuated equilibrium." There are long periods of slow change in the evolution of species, and then there are times of sudden, rapid change, like an asteroid strike, or sea level rise. This drastic shift in the environment prompts a massive bloom of change in many species, some of which end up better adapted to the new environment. Let biology teach us that we cannot go 110% all the time. We bide our time, and occasionally, when we see an opportunity, we jump and commit the extra effort. Exert 110% at the right moment.

Look for moments to apply punctuated equilibrium in your career. Has there been a change in leadership at your organization, creating a new chance to show your worth? Is your organization adopting a new technology? Being one of the first experts in the new tech may earn you greater recognition.

But remember, choose carefully. Don't jump at everything. Such behavior leads to exhaustion. Realize that every career has both leaps and lulls. Just because someone else is leaping doesn't mean that you have to.

The Lesson Learned

In an arena of high-performing individuals, you will constantly see evidence of success all around you. Be forgiving when you see others being successful and don't chastise yourself. Your circumstances are different. Self-forgiveness is an effective strategy; so is quiet vigilance. There are always opportunities for punctuated equilibrium—times when going all-in may create an advantage. Save your energy, look for these chances, and then pounce.

The Startup CEO's Little Black Book

According to my smart younger sister, all the cool west-coast startup CEOs carry a black book and wear sneakers and blazers over ironic T-shirts. I figure I could at least carry a black book. The book and my smartphone go everywhere with me. What's in it? What's it good for?

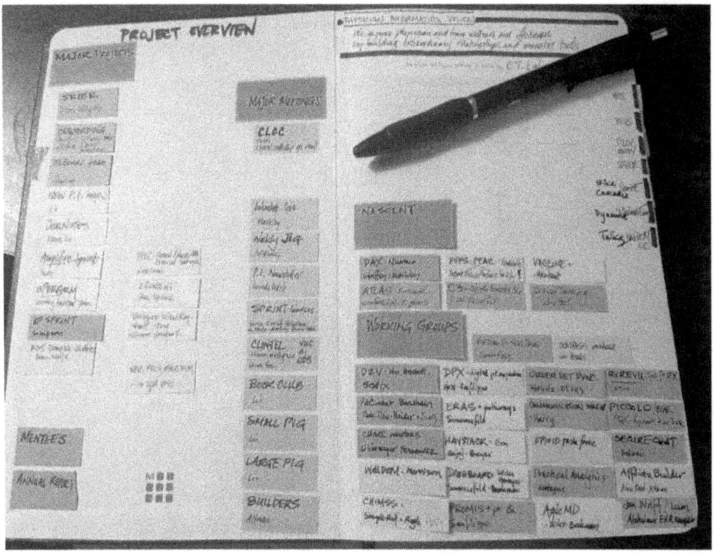

Here it is. I show you my black book to get your own creative gears turning.

It is...

- An indicator of your presence and attention: It is clear when you take notes that you're not answering texts or emails or doing other work.
- A tool for better recall: Since handwriting is slower than typing, it forces the brain to assimilate what it is hearing as it writes. Even if I never read my own notes, I still remember better.
- A tracker for long-term projects: You can assign dedicated pages to special people, ideas, and projects.

- Totally wireless: No dependency on electricity, Wi-Fi, tablets, or laptops.

My black book is not just a place to record my thoughts; it is also a personal planner. I use three-inch square sticky notes, cut to size, to index the book (see the sharpie marks on the right edge), and I use my opening page for major projects so I can absorb, at a glance, what my team and I are working on. (The other thing it shows me is that I have too many priorities. The struggle. Is real.)

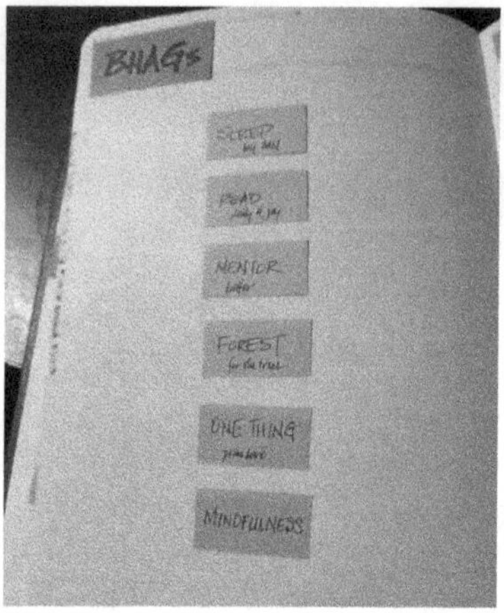

I use another page to list my **BHAGs: Big, Hairy, Audacious Goals**. The idea here is that setting small goals and achieving them doesn't inspire dramatic action. For that, you need a BHAG. This concept was lifted from *Great by Choice* (by Jim Collins and Morten T. Hansen).[1] Some of my big goals are hairier and more audacious than others—someday, some year, I will sleep eight hours a night. I do read regularly. I enjoy mentoring and want to keep getting better.

I like seeing the forest beyond the trees, when I remember to do so. I constantly remind myself to prune down my priorities to *one thing* and use the Pomodoro technique to manage my time (see **The Short View: Pomodoro Timers & Email Tricks** on page 291). And who couldn't use a dose of mindfulness throughout the day?

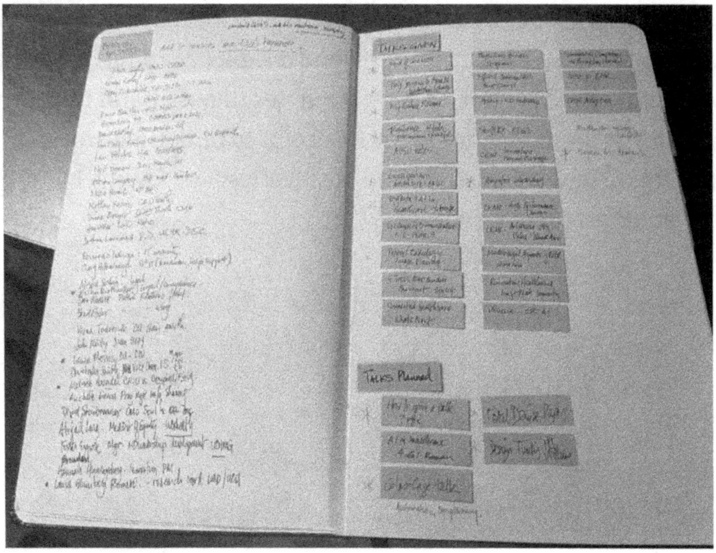

I have a terrible memory for names, so as I meet new people, I record their names in my black book. I also track the talks that I have given and the ones I'm thinking about or planning.

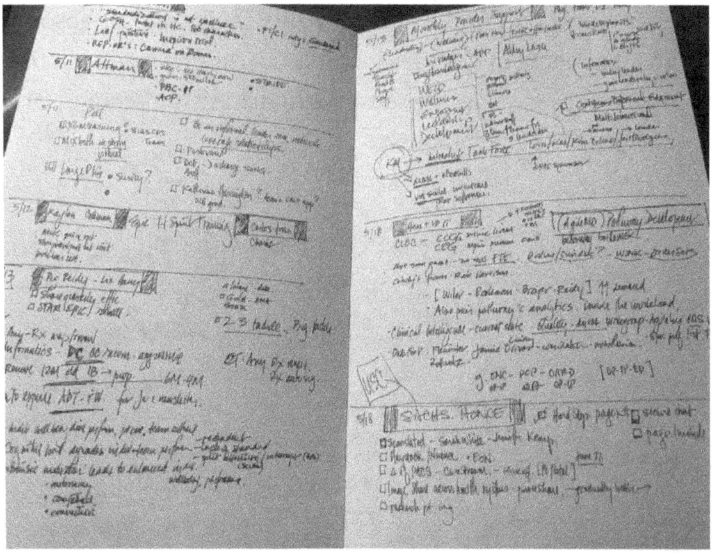

The rest of my book contains actual jotted notes taken during meetings, including major topics, my tasks and insights. My direct reports each get a bookmarked section to keep my notes together and improve my continuity of thought. I use a contrast color to underline any to-dos or high points to remember. At the end of a meeting, I review my highlighted to-dos and start them right away.

If I take my book on trips, I'll find time to sit and sketch. I sketch from a picture displayed on my propped-up phone, at the desk in my hotel room, just before bed. I find sketching to be relaxing and all-absorbing. It is a moment of being in flow. The point is not the end result but the time spent totally focused, using the senses and creating something. Not all my sketches are great. For example:

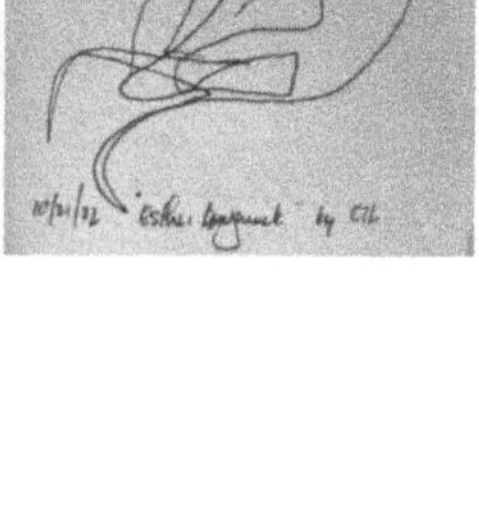

Hopefully my own clumsy attempts will inspire you to try it! Spend an hour, don't be too judgy on yourself, and enjoy the process. Did the hour pass quickly? Did your monkey-mind-chatter subside? Sketch or doodle when you can, and you might surprise yourself. I do sometimes.

The Seattle Public Library, a feat of architecture and engineering and art. You'll note that, for me, straight lines: easy; human bodies: hard.

Once you have the habit of carrying your little black book, the possibilities are endless. I've written "three good things" as a daily reflection on gratitude. I've also recorded "one thing I learned today." In short, there is no end to the interesting things you can scribble in your book.

My preference is for paper with no lines, so I can sketch anytime. Other black book owners may prefer a grid or lined paper. Whatever works for you is what's right. I also love going to stationary stores to buy Kurutoga pencils, rare gel pens (JetPens.com, anyone?), new moleskin books, stickers, and washi tape.

Finally, as mentioned above, I try to avoid taking notes on my laptop or phone. Typing on your laptop or phone is often perceived unfavorably by those around you. You look like you're mentally absent, doing some other work, when you could be paying attention.

Also, research shows that when others in an audience can see your screen, not only is your screen decreasing your comprehension of the lecture, session, or meeting, it is decreasing the comprehension of *everyone who can see your screen*, even if they can't see it clearly.[2] (Meeting-laptop-users, *you* are the problem.)

I know, electronic note taking allows for easy searching and indexing, and there are Rocketbooks and other hybrid devices for people who want the best of both worlds. I have not found success with them.

Besides, it's much cooler to use an old-school pen and paper. I love the tactile nature of paper, the expressive personality of handwriting, the ease of drawing lines and sketches to link ideas.

> **The Lesson Learned**
>
> Physical writing is slow and enjoyable, and it makes you think. I show you my examples, but a book with blank pages is a wide-open invitation to make it your own. If you want to take notes that can be searched and indexed, do it electronically. If you want to write, reflect and think, take a page from the startup CEOs: Carry a black book.

Getting Things Done in an Age of Acceleration

We are all drowning in minutiae.

For every email we answer, for every small problem we solve for a colleague, we get a small dopamine hit of satisfaction. Hit that send button! Ahhh, yes.

However, after a long day, we think: "What did I do today?" And we realize, "I was in my email for hours. I replied to a lot of people. But I didn't make a bit of progress on that Big Thing I want to accomplish." In this subsection of Thrive are specific tactics I find helpful in the swirl of daily life.

The Sisyphus Boulder and the Eisenhower Box

In Greek mythology, Sisyphus is a Titan deity whom Zeus dethroned from Mount Olympus. For his punishment, he was made to push an enormous boulder up a hill for the rest of eternity. As soon as the boulder neared the top, it would roll back down again, and he would have to start over.

Today, we are all Sisyphus.

It wasn't always this way. Messages used to come in the form of scribbled words on paper in a manila interoffice envelope tied with red string. We would read the paper, handwrite a reply, put it in the envelope and send it back to be delivered in a few days, or if no reply was needed? Into the trashcan.

In 1995, when I was Physician Liaison to the IT department, email was still new. Every day, around lunch time, curious if I had received any electronic messages, I would sit at the computer, launch the telephone modem, listen to the modem dial the phone and connect, then with bated breath, run my email program. Most days, my email inbox was empty. I received a new email every few days. When an email arrived, it was exciting!

Things have changed. I now lead a group of 25 and interact with an IT team of over 300 working on dozens of active projects within an organization of over 6,000 physicians/APPs. I receive over 250 emails per day. I have long since turned off my "You've got mail!" alert.

Inbox Zero is a tactic from David Allen's *Getting Things Done*.[1] The idea is to empty your inbox of messages at the end of every day, thus allowing you to head home with a clear desktop and nothing hanging over your head for later. In our accelerating world, however, Inbox Zero should not be your goal. I tried it, spending ever more time digging to get to zero before going home. I would miss dinner.

I have read numerous books on work efficiency and tried various techniques. I now recognize the overwhelming volume of daily messages. So, what is a person in a leadership position to DO?

Let's start with some DON'Ts.

Don't pursue Inbox Zero. My colleague blew my mind when she asked me, "Is replying to email your primary value to your organization?" I had to think, and my response was an embarrassed "Um. No."

I'm here to organize teams to achieve strategic goals. Replying to emails builds connection and speeds up some communication tasks, but it is not my primary job. From the above example of 250 emails a day, if I took an average of two minutes to respond to each email, that's 500 minutes. An 8-hour day comprises 8 x 60 = 480 minutes. The math does not look good. This question triggered a paradigm shift for me. Be selective. More on this later.

Don't multitask. You may believe that you're more efficient doing two things at once. You are wrong. Cal Newport's book *Deep Work: Rules for Focused Success in a Distracted World*[2] emphasizes the power of disconnected, uninterrupted time. Nicholas Carr's *The Shallows*[3] speaks of the opposite trend—that constant internet access promotes shallow, superficial thinking and makes it much more difficult to do focused work. It is so easy to look one thing up, click the next link, and the next, and soon you're looking at cute kitten videos instead of finding an important fact for a project you thought you were working on.

One of the biggest takeaways from Newport's work, for me, was "attention residue." Switching rapidly from one task to another causes focus to drop for up to 15 minutes. The split attention leaves a "residue" with parts of your brain still thinking about other tasks; there is less brain power left to work on your complex, important primary task. In other words, effective multitasking is a myth.

Don't save your email for evenings and weekends. This one is my personal Achilles' heel. I was trained in the era that taught "evenings and weekends are when you get ahead as an assistant professor, if you want a promotion." This voice constantly whispers in my ear, telling me to just do a *little* work on Friday night.

In years past, if you had a question, you had to get up from your chair. You had to walk over and knock on someone's door. Or make an appointment to meet. Or handwrite a memo and send it by interoffice courier. The effort of sending a message limited the number of messages sent. That friction is gone.

With smartphones and constant access to messages, work can and does expand to fill all possible time. It takes seconds to have a thought and send a message, and far longer to compose a thoughtful reply. There is an impossible asymmetry between sender and receiver.

It is important to have compassion for yourself. You are not a reply-robot.

One of my colleagues recently got divorced, and his work was a contributing factor. For years, he had been amazingly responsive to email, even replying outside work hours. It is so easy to be "always on" for work, but it can quickly unbalance our personal lives.

Don't use your inbox as your to-do list. Upon learning I had this habit, another of my colleagues told me: "No, your email inbox is someone else's to-do list for *you*." Another mind-blowing thought. Could it be, replying to email is for suckers? What, then, should I do with all these plaintive requests for information, help, or advice?

On a bad day, I am sucked into all my bad traps: opening my email first thing, replying to emails in reverse chronological order, until I look up at the clock, and ack! The morning is gone, and I'm late to my next meeting.

Do away with these habits—instead, set a BHAG. The idea of having Big, Hairy Audacious Goals was coined by Jim Collins in *Good to Great*.[4] I like to discuss and establish our BHAGs within our informatics group for the year.

For example, previous BHAGs for the EHR at our organization have included:

- Convert all progress notes at our organization from SOAP to APSO format.
- Be the first large organization in Colorado to make OpenNotes the standard of care.
- Reinvent ambulatory EHR optimization by creating Sprint teams.
- Implement ambient notes for all providers at our organization (my current BHAG).

Taking time to identify Big, Hairy, Audacious Goals is an important exercise. Those BHAGs serve as clear beacons of what you intend to achieve. For better or worse, they also serve as reminders of what you are *not* achieving when you're embroiled in day-to-day work putting out fires, responding to emails, requests, and In Basket messages.

BHAGs are not urgent; there are no deadlines except the ones you impose on yourself. Do not mistake your busyness for taking care of business. Unlike our emails, chats, and texts that "ping" and clamor for our attention, non-urgent BHAGs sit silently at the edge of our perception. If we are not careful, days, weeks and months pass with no progress on important work. I have learned from the many long-term projects that failed, and the few that survived. Creating habits and protecting time to focus on *one thing*, one BHAG, is crucial to success.

The Eisenhower Box

The Eisenhower Box, supposedly created by Dwight himself, is a good way to think about BHAGs. The Eisenhower Box is a 2 × 2 table like the one below.

	Urgent	Not Urgent
Not important	Daily emails	Safe to ignore
Important	Crisis. Just do it	Strategic projects

One axis is labeled "Urgent" vs. "Not Urgent." The other axis is labeled "Important" vs. "Not Important." This creates four quadrants that represent the four types of tasks we encounter. Responding to the Urgent and Important box is easy: Drop everything and do it. One example would be the medical crises at the beginning of the COVID pandemic. The non-urgent, non-important is also easy: it's spam. Don't do it.

The other two boxes are harder.

Email (mostly) goes in the "Urgent but Not Important" box. On the other hand, strategically important projects land in the "Important but Not Urgent" box. If you do respond to what is Urgent, then unimportant tasks will fill your day. You will never find time to work on what is Important.

> **The Lesson Learned**
>
> Don't let email run (and ruin) your day! Set a BHAG. Use the Eisenhower Box. Stop the tyranny of the urgent and build habits to work on non-urgent but strategically important projects. Do not mistake your busyness for taking care of business.

Get in "The Zone"

When do you do your best work? When are you "in the zone"? What activities can you disappear into? What work do you most look forward to?

Mihaly Csikszentmihalyi wrote about this in *Flow: The Psychology of Optimal Experience.*[1] He defines "flow" as: "...the state in which people are so involved in an activity that nothing else seems to matter; the experience itself is so enjoyable that people will do it even at great cost, for the sheer sake of doing it."

We have all felt this at one time or another. Flow surprises me: It doesn't occur during passive activities, like watching TV or reading. Flow is only achieved when the task is challenging, requiring skill and concentration, when there are clear goals, and when you get immediate feedback on your actions. Only then can you experience a deep, effortless involvement, a sense of control. Your sense of self vanishes and time seems to stop.

My common flow states occur in challenging meetings at work, when we snatch a win-win agreement from the jaws of defeat, or when working with a patient on a difficult behavioral change and achieving a breakthrough in their health.

I am also lucky to achieve flow states outside my work. Literally! Colorado is a wonderful place for cyclists. During one such ride, my colleague and I summited the climb up Loveland Pass at 12,000 feet. As we descended the broad, sweeping curves at 40 miles per hour, every bit of my concentration was focused on balance, anticipating traffic, braking and accelerating at just the right moments. My sense of self vanished. Everything became the task at the edge of my skill. It was a blast.

The Pulitzer-Prize-winning biographer Ron Chernow describes Alexander Hamilton writing the Federalist Papers: 55 papers, all written in a few short months.[2] The volume of writing is astounding. What's more astounding is that Hamilton accomplished this writing in his evenings, *after* a full day in a busy law practice. He would spend hours consumed in reading. Then he would sit down late at night and write a completed paper in one frenzied sitting. Hamilton had clearly found his "zone."

The specific ingredients of your own flow state may be different, but they will likely comprise a challenge, your enthusiasm, and your own skill. Build time into your week to maximize "in-the-zone" time for yourself. The habit increases your productivity, yes, but—and more importantly—it increases your joy.

Achieving flow can also be approached in reverse. Sometimes your tasks are too difficult to achieve the "flow" at the edge of your skill. Do you need to grow certain skills to match your challenges and to achieve flow more often? For example, do you need to learn to run meetings more effectively? Do you need to practice writing story-focused one-pagers to improve your communication? Do you need to read about designing better talks to be an effective presenter? These topics are all covered in these pages and elsewhere. Select projects and/or grow your skills to maximize your time "in the zone."

Even one or two weekly time blocks allocated to a flow-state activity can significantly improve your wellbeing and reduce burnout. Achieving flow gives your work meaning. It lets you feel like you are being stretched, and that your skills are being used and appreciated. It helps you grow in surprising ways.

The Lesson Learned

Do you know the conditions that allow you to become lost in your work? To boost productivity and joy, find time every week in your schedule to be "in the zone" working in a flow state. Remember the three components: your skills, concentrated on a challenging but achievable task, toward a clear goal.

Return to Your Breath

One of the benefits of mindfulness and meditation is improved focus. Let's say you've blocked out two hours to dive deeply into a project. The scheduled time arrives. You've cleared your desk, turned off your internet, laid out your pen and papers. But before you know it, you realize you are thinking about something else. Darn! Okay. Take a minute to just breathe. Focus. There is your work in front of you. Although—hang on, what was that thing you have to do after lunch...?

Staying focused is hard work. It is particularly difficult when social media and smartphone apps and games are built to capture and keep your attention as often as possible. Many of us work in open cubicles where anyone can walk by and tap our shoulder. Some of us are required to keep constant access to an instant message app while at work. Focus seems impossible.

One technological way to fight back against this mind-chatter, and the distractions of the "attention economy," is to shut off all non-critical alerts. Be ruthless at removing every "ping" and "buzz" that you can. All phone apps and email clients have "focus" settings or a way to mute notifications. Set them up.

More important, however, is to address the chatter that goes on in your mind.

You can start a mindfulness practice with no technology at all. Find a quiet spot to sit comfortably where you will not be interrupted. This might be a bench outside your building. Or an unused conference room that is empty of people, phone, and laptop. Or on a slow walk along a less-traveled path. Or even at your own desk, if you can really block *all* interruptions for a short time.

Close your eyes and focus on the sensation of breath at the tip of your nose as you slowly breathe in and out. You can close your eyes or keep them partly open. Begin with a timer for two minutes. Any time you notice your attention wandering, label it "sound of motorcycle" or "air condition sound" or "yesterday's argument" and then return to your breath. Just being aware of how much chatter your monkey-mind generates is a huge insight, and repetitive sessions of "returning to breath" can improve your ability to focus for the rest of your day. Don't be surprised if you detect more than a dozen interruptions in two minutes. This is hard work. Consider every detection, not a failure, but a success. Noticing and returning is the goal.

If you need extra help, there are mindfulness apps. Seemingly ironic, these apps can be your gateway to mindfulness. Informaticists sometimes enjoy experimenting with technology if it helps achieve their goals. I like the Apple Watch mindfulness app. I have also experimented with Headspace,[1] Heartmath,[2] and Muse.[3]

The Lesson Learned

By developing habits around flow and focus, you are practicing and improving your mindfulness. Longer and more frequent periods of flow and focus are shown to be associated with greater happiness. If you are feeling burnout, this is one treatment.

The Long View: Leading Indicators & the Power of Habit

Don't use lagging indicators as your metric of success. For example, "number of research papers published" is a lagging indicator. Lagging indicators measure the outcomes of all the work that went *before* them, and they are poor motivation. "Why haven't I published more?" is just beating your head on the wall.

If you were planning on running your first marathon, a lagging indicator would be like asking yourself, "How many marathons have I run?" Zero. And the next day: still zero. Ask yourself this question every day, nothing changes, and your motivation wanes.

Instead, go after *leading indicators*. These are measures that move you toward downstream goals. For example, I used indicators like: "Write for 30 minutes every morning," or "Don't look at your email in the morning until you have done one hour of writing." Try out some leading indicators of your own until something works, make note of them somewhere highly visible, and build a habit around them.

For a marathon planner, a leading indicator would be "Did I run a mile today? Yes." Then, after routinely achieving that goal for a few weeks, setting progressively more difficult intermediate goals that can be achieved toward your ultimate goal.

If the structure is working for you, it can be helpful to have nested sets of indicators. Your daily plan: Use a pomodoro timer to manage your hours. (More on this below.) Weekly plan: Block out time each week so it adds up to a planned goal.

Thomas M. Sterner describes this framework in *The Practicing Mind: Bringing Discipline and Focus into Your Life.*[1] What seems to be an entertaining book about practicing music and sports is *actually* about mindfulness, happiness, and living your life fully.

Sterner draws a contrast between the experiences of young and older learners. The young mind, being told to "practice for 30 minutes," finds it boring and depressing. The young person can only dream of the out-of-reach skills they would like to have. Sterner compares this to an adult learner with more life experience who is practicing music for enjoyment. In the second case, the practice itself is joyful: "I am constantly getting better, even this minute."

This difference is also shown through the metaphor of one sailor who is always staring at the horizon, toward his destination. He finds that that the horizon never changes and is constantly frustrated at his lack of visible progress. A second sailor instead looks down at the waves crashing against the bow and the dolphins that pass through the ship's wake. This second sailor is aware of and enjoying the present and the progress she is making.

The idea of lagging and leading indicators expresses this same comparison: Instead of asking yourself, "Have I done the thing yet?" find ways to measure the progress you're making toward that goal. Design your own leading indicators and then design your days and weeks to achieve them.

Of course, no matter how carefully you plan your days and weeks, you won't see progress on your leading indicators if you don't build them into habits. We all have bad habits we're trying to quit and good habits we aspire to. But why is it so hard to engineer our brains and bodies to do the things we know we *ought* to do?

In *The Power of Habit*,[2] Charles Duhigg identifies the three factors of Cue, Routine, and Reward. In brief, you must recognize your cues and cravings, then design situations and rewards to extinguish your bad habits and build desirable new ones.

For example, say you're trying to establish a morning jogging habit, but you can never get yourself out of bed. **Cue**: Recognize that you don't like going out in the cold morning air. **Routine**: Lay out your two layers of jogging clothes at the bedside the night before so they are the first thing you see. Finally, set yourself a daily **reward**, like a checkmark on a calendar, with a promise to buy yourself a small gift when you complete X number of morning jogs. Do this for 30 days, while the habit is at greatest risk of fading. Once you've completed 30 checkmarks, it can become self-reinforcing as you start to see and feel its benefits.

The Lesson Learned

If you only focus on a distant goal, you will never see the progress along the way. But if you don't set yourself up for habitual success, you'll struggle to make progress in the first place. These two long-term strategies, leading indicators and the power of habit, can together set a pattern of progress for any work project and any self-improvement goal. Pick something small and get started!

The Medium View: Accountability & Tuesdays to Write

Starting in 2003, I used the Dragon speech recognition program to dictate my internal medicine progress notes into the EHR. I typically did this at the end of the day, but in 2005, I had the great idea to dictate my assessment and plan in the exam room, in front of the patient. Then I would print out the plan for the patient to take home. My patients loved it.

I hadn't heard of anyone else applying Dragon this way, and I thought to write a paper about "Speech Recognition in the Exam Room." Unfortunately, this idea got buried in the drawer of good ideas, as my attention was pulled elsewhere.

About 20 years later, while meeting with a physician informatics colleague about a completely unrelated matter, I brought up the idea of "speech recognition in the exam room", including that I wanted to write a paper on the idea but had been struggling to find time. He was excited.

"It's a great idea," he replied. "I'll help you do it if you help me write it."

An enthusiastic partner is a precious resource. I rarely turn down an opportunity to collaborate with a colleague who shares a mutual interest. In my personal life, I recognize that having a friend with a common interest increases my chances of achieving a goal. For example, when attending karate classes, or completing a bike-packing trip through the Cascade Range in Washington State.

After a brief discussion, we agreed to meet for one hour twice a month. Over the following six months, our accountability partnership spurred steady progress even despite our busy careers. We designed a survey for patients of doctors who had used speech recognition in the exam room. Then we distributed our survey, collected the data, and analyzed the responses. We described our work and submitted our manuscript to an informatics journal, handled the inevitable ego-crushing reviewer comments, and finally got our manuscript accepted.

Organizing, conducting, and publishing a study was much easier with an accountability partner. One thing that really made a difference was having a consistent block of time to devote to the project. We had mutually agreed to hold each other accountable to keeping that time. But can you create that accountability with yourself?

The idea of "Tuesdays to write" came from a colleague of mine who blocked out his every other Tuesday morning from 7:00 a.m. until 1:00 p.m., during which time he turned off email notifications and disabled his access to the internet.[1] He planned his Tuesday morning ahead of time and got right to work on his non-urgent but important project (see **Getting Things Done in an Age of Acceleration** on page 275). Over time, his colleagues came to understand he was never available on Tuesday morning. Sounds blissful, right?

Set aside time in your calendar. Consider it "making an appointment with your working self," during which you'll close your door or put on your noise-canceling headphones and turn off notifications. Decide on your *most important* project that week and then spend at least two hours assembling information and building something you'll be proud of. You don't have to finish it; you just have to dedicate a solid chunk of uninterrupted time to the task.

Think "process," not "product." If you're stuck trying to get started, tell yourself you'll work on the first step for 30 minutes. Maybe that's just tracking down the phone number of the person who might have the answer you need. Maybe that's putting down a terrible first draft of a document you promised to write. Whatever it is, just start.

In *The Practicing Mind: Bringing Discipline and Focus into Your Life*,[2] author Thomas M. Sterner outlines a framework for being present and chunking work into digestible segments. He refers to this as the "Four S-words"—Simple, Small, Short, and Slow. For example, let's take the dreaded task of cleaning out the garage:

Simple: "I will do exactly these three things, and no more."

Small: "I will only clean from the north corner to the left until the window."

Short: "I will only spend 45 minutes on this task."

Slow: "I will be aware of every movement and action that I am taking."

Turning big, unpleasant jobs into simple, small, short, or slow tasks is an effective way of handling an initial lack of inertia. It can be a jumpstart for the hours of "Tuesdays to write."

The Lesson Learned

On a basic level, productivity is about ensuring you have the time to work and *use it*. If necessary, find a partner to keep you accountable. Schedule time every week to work on important, non-urgent projects. This is the essence of "Tuesdays to write."

The Short View: Pomodoro Timers & Email Tricks

Sometimes, the focus just isn't there. Especially if you are tired, hungry, unsettled, or excited about your weekend plans, it can be tough to knuckle down for a marathon work session. If the marathon isn't happening, how about a 100-meter dash?

Don't struggle to keep your focus indefinitely. Consider the Pomodoro Technique by Francesco Cirillo. *Pomodoro* is Italian for "tomato," after the tomato-shaped timer Cirillo used when developing the method.[1] Cirillo's *pomodoro* measured 25 minutes, and the method was simple: Set a timer, work until it rings. When your time is up, you stand up and stretch. Setting a timer, working for 25 minutes, and then taking a five-minute break helps create discipline in short bursts, and the ticking timer is a reassuring reminder that you're on task.

Cirillo recommends keeping a yellow pad and pencil nearby as you work. There *will* be stray thoughts—tasks you suddenly remember, groceries to be purchased, emails to send. These distractions interrupt your productivity. Write them down so they can live somewhere outside of your short-term memory and then *return to your task.*

If you have time for several tomatoes in a row, take out a piece of paper and draw a stack of checkboxes. Next to each, jot a task that you can do in 25 minutes (e.g., "Write outline of the proposal" or "Call your patient about that test result"). Cirillo recommends up to four tomato timers with five-minute breaks, after which there is a longer break, for a total of 2.5 hours. You can then repeat this 2.5 hour cycle all day.

At the end of your time, look back at your planned list and compare what you did with what you *thought* you would get done. Over time, you will get better at predicting what you can do in 25 minutes, and then how many tomatoes it will take to do something. You'll feel less frustrated as you get to know yourself better.

The five-minute break between 25-minute bursts of effort is also purposeful. It allows both a mental break and some unfocused time for the brain, which is important to consolidate memory,[2] and allow lateral thinking by your subconscious.

Furthermore, the quiet, calming, yet insistent tick of the timer reminds you to stay on task. The tick tells you, "Don't worry, you're doing great. Keep going."

You can buy a *pomodoro* timer—tomato-shaped or not—from any kitchen store. If you prefer a high-tech solution, there are smartphone apps. I like one called "Focus Keeper."[3]

Taming Your Inbox

The email inbox is a villain. It robs our time, invades our reflective moments, and follows us home at night. It takes cunning tactics to outwit the inbox and turn it from a traitor back into the tool it was meant to be. Here, in no particular order, are some tricks you can apply to your own inbox.

Crush the spam. Learn to mark unsolicited emails as spam. Take the time to unsubscribe from newsletters. You'll quickly see a noticeable difference. Remember to occasionally comb through your spam folder to spot desirable emails that were mistakenly categorized.

Group your inbox by conversation. Temporarily sort by "oldest first," then delete entire conversations that are no longer relevant (very satisfying).

Forward emails to yourself on a specified date. Or, move emails into a "work later" folder. Then make sure you've blocked time later in the week to handle that email work. Respond in decreasing priority. Select how many pomodoro timers you will spend on email. When your time is up, just stop. You *can* put a boundary on boundless work like this.

Create a folder and filter for "high-priority" colleagues. When you next come to your email, look at the new messages in your priority folders first, before looking at your general inbox. For example, my boss gets a priority folder, and so does my Physician Informatics Group members (Large PIGs). Everyone else is relegated to the general inbox, where they may (or may not) eventually be seen. I'm sorry, okay? We are all doing the best we can.

If necessary, declare email bankruptcy. In an extreme case, group-delete anything from more than six months ago. If it is July right now, are emails from January still relevant? Delete them. Their senders will always find you, if it really matters.

Use the subject line like a text message. If you don't have much to say, sometimes a short blurb in the subject line is enough. Leave the email body blank. For example: *SUBJECT: Should we cancel next Tuesday 11am? I have nothing new. CT*

Ending with your name also signals there's nothing in the body. This SMS text-like brevity is quick and easy to read and allows a quick and easy reply.

FYI vs REPLY REQUESTED. The subject line has other utilities as well. One trick I learned from my boss is to use the subject line to indicate whether the email can be read without responding. This is a kind thing to offer your busy colleagues who may also adopt the habit).

FYI ONLY: New classification to use next time

Vs.

REPLY REQUESTED BY EOD Wed 3/25: Need to know if you're attending

There are tons of other email tricks. Keep learning, read an article, ask a colleague, try a new efficiency tip from time to time.

The Lesson Learned

The pomodoro technique is a handy way to chunk your time and stay focused. Also, help yourself and your close colleagues by using email effectively.

Scrolling Alone Vs. Laughing in Groups

I participate in the wellness initiatives at my organization. As such, I support the idea of EHR Optimization Sprints (see **Sprint Informatics** on page 105), to boost physician and team efficiency with the electronic health record. However, we know that physician burnout and personal wellness are affected by many things, as described in the Stanford Wellness Model.[1]

Described by Tait Shanafelt, MD, it posits that professional fulfillment has three aspects: a culture of wellness, personal resilience, and efficiency of practice.

Although much of my work as a CMIO is dedicated to **efficiency of practice** with respect to the electronic health record, our Large PIG (see **What Is Your Vision Statement?** on page 137) and Sprint teams are well aware that in treating burnout among our clinical colleagues, increasing efficiency is not enough.

Indeed, improving efficiency of practice is great for those who are receptive and ready to learn the skills. However, quite a few colleagues suffer from deficits in their working environment—they don't inhabit **a culture of wellness**. In brief, a clinician's professional fulfillment is highly influenced by the behavior of their local leaders: the clinic medical director, hospital operations leader, and the various managers who set appointment durations and vacation leave policy, support paid time off, and respect the autonomy of their staff.

Others suffer, after years working in a difficult environment, from a deficit in **personal resilience**. This is not to accuse those who survived a decade of intense medical training of "lack of resilience." Regardless of personal fortitude, one's will can be slowly whittled down over years in a chaotic work environment. This chaos results when physician-nurse-clerk teams are frequently shuffled. Newly shuffled teams don't work as efficiently together and can't anticipate each other's needs.

Another source of chaos can be constantly shifting expectations about productivity. A physician may be asked to reduce the length of their patient appointments, or to see more patients per day. Patient portal message volume ballooned by 350% during the COVID pandemic, and responding to those messages turned into after-hours pajama time with the EHR, impacting family and personal life.

In thinking more about these interlocking influences, I'm reminded of the work of Robert D. Putnam, who wrote *Bowling Alone*.[2] This insightful work documented the decline of civic engagement in the United States. The title of the book refers to the fact that membership in bowling leagues declined 40% between 1980 and 1993, while the number of individual bowlers rose by 10%. This is an illustration of the dramatic decline of American participation in social organizations outside work, and it demonstrates that recreation has increasingly become a solo activity. Bowling was just the illustration, showing how face-to-face social gatherings experienced a dramatic decline over the past decades.

Putnam proposed, in 1995, that this unraveling of the social fabric of American society led to isolation, loneliness, and a corresponding decline in civility and personal resilience. Of course, loneliness and incivility are even more apparent now. The dramatic increase in social media use—the thin veneer of reading semi-anonymous social media posts (many of them written by bots)[3] increasingly displaces deep in-person connections with real humans.

The Department of Medicine runs a group composed of clinicians volunteering as "wellness champions," who seek out and publicize resources to improve wellness, efficiency, and resilience among our faculty and staff. Our wellness champions recently read a book called *Together: The Healing Power of Human Connection in a Sometimes Lonely World*, by Dr. Vivek Murthy,[4] former surgeon general of the United States. He calls attention to the epidemic of loneliness and how we are all suffering from its effects, both in our families and in our communities.

There are counterpoint arguments to this assertion. The 1996 *Atlantic* article "Kicking in Groups" by Nicholas Lemann[5] argues that the new social capital resides with informal groups like soccer leagues. Putnam's original data is too selective, it asserts, and the rise of new dynamic social groups offsets this trend. Lemann argues that people are out there "kicking in groups" in soccer leagues, but we don't see it in the data.

Reading the debate more than two decades later, I'm more on the side of Putnam than Lemann. I am seeing anxiety and loneliness among physician/APP colleagues as well as in many of my internal medicine patients. "Bowling Alone" and "Kicking in Groups" are old terms from the '90s. We need new metaphors. I like "Scrolling Alone" versus "Laughing in Groups."

As informaticists and leaders, what should we do? What *can* we do?

Remember the original Star Trek? Captain Kirk often turns to "Bones," the ship's doctor, and asks for unreasonable cures or fixes, prompting Bones to reply, "Dammit, Jim! I'm a doctor, not a bricklayer!" In this case: "Dammit, Jim! I'm an informaticist, not a social scientist!" The loneliness problem is a massive one, and we are all unqualified to fix it. Nevertheless, we must start somewhere.

One thing we can do is figure out how to measure it. We already see in our survey data with our physician/APP colleagues that burnout is high, at 40–50% in national surveys. "Intent to leave or cut back medical practice" in the next few years is even higher (90% in some internal surveys).

It is time we started asking about loneliness. I believe it is an important underlying factor for burnout and "intent to leave."

It may be time to ask yourself, do you have a social group that you meet with at least monthly at work? What about a social group that you meet with at least monthly *outside* of work? Do you meet regularly with a mentor or mentee?

I believe that measuring such behaviors is a way to determine who is vulnerable to burnout and who is protected. If loneliness is a contributor to burnout, consider the following ideas:

- Create a goal that everyone has both a mentor and mentee relationship and meet regularly.
- Social gatherings regularly hosted by clinics or groups.
- Balint groups that meet monthly (see below).
- Running a Wellness at the Table program (see below).

A Balint group,[6] comprising physicians and advanced practice professionals, meets monthly to present clinical cases and seek to better understand and improve the clinician-patient relationship. While clinicians are usually trained to seek the right answer to clinical problems, in a Balint group, the focus is on enhancing the clinician's ability to connect with and care for the patient. Such groups become tightly knit and are an effective social connection and bulwark against burnout.

Wellness at the Table[7] is a UCHealth program that funds up to $25 per attendee to meet at a restaurant, have a meal together, spend at least 15 minutes talking about a wellness topic, and then enjoy each other's company outside the pressures of direct patient care.

Rather than wait around for social movements to get started, why don't we start something where we are? Measure the loneliness. Talk about it. Encourage social interaction, recover our lost habits of community building. We were already sliding, but the pandemic threw us over the cliff. It is time to claw our way back and rebuild our communities.

The Lesson Learned

Loneliness is a major component of burnout. If you can measure loneliness, you can see the extent of the problem and work on solutions. Two concrete actions to take: encouraging mentor/mentee relationships and social groups like Balint. Let's encourage Laughing in Groups.

Write like Hell

At the beginning of my career as a physician, my boss told me: "Your clinic is where you see and treat patients. It is also where you teach medical students and residents. It is also your laboratory for academic study." We were taught, when we arrived as junior faculty at an academic medical center, that we were expected to be "triple threats"—excellent clinicians, excellent teachers, and excellent researchers who advance the science of medicine.

Whew. That was a lot to stack on the shoulders of a 28-year-old who had foregone his young adulthood to study suffering and disease and make this first step in his fledgling career. Nonetheless, I was optimistic.

Seeing patients—that was easy. Come to the clinic or to the hospital, see patients as they come to you. Do your best job, learn from consultants. Read like hell. Continually improve your knowledge and your skills.

Teaching—that was easy. Accept the students and residents who rotate onto your service or come shadow you in clinic, give them gradually expanding responsibilities as they learn to do what you do. With feedback from students and peers, learn to be a better teacher over time. Read like hell.

Publishing papers—that was hard.

Unlike teaching and clinical care, conceiving a research or quality improvement study was really difficult. I asked my boss: "When will I have time to do this? I already take my charts home to write or dictate my notes. I already stay late at the office to return phone calls after hours."

His reply was simple and unhelpful: "You have every evening and every weekend." This glib advice mirrored the standard expectation for all junior faculty. (Burnout in academic medicine did not start with the electronic health record.)

Leaving aside the unreasonable expectations from my boss, the bigger lesson stuck: Our work, whether in clinical medicine, or in clinical informatics, is incredibly rich soil for growing the tree of knowledge. However, with my lack of research experience and skill, instead of a tree, I would be lucky to grow a sickly shrubbery.

Over the years, I have set myself the goal of looking at *every informatics project* as an opportunity to present at AMIA, AMDIS, HIMSS, or at one of Epic's national meetings, or to write and publish an academic paper. I have learned that it doesn't take major funding, or a PhD in statistics, or a large research team. My paper-writing team can be a team of one. However, it's more *fun* if you have a colleague who is willing to put in the hours and coauthor a paper or presentation.

The beginning is always difficult: What's a good question to ask? What's a good project to study? Hasn't someone already written about this? Isn't this a waste of time? What if we get rejected? And ugh, I don't have any time right now.

All of this. All the doubts ready to stop us. And yet, if there is a will, there is a way.

In our group, we have no statistician, but we know some nearby. We have no manuscript writer, but we know some medical writers who are willing to edit our work. We have no time protected for research or quality improvement. We are a bunch of physician/APP informaticists who have a shared interest, who are willing to carve out an hour or two per week, and *just write something*.

I think back to the Practicing Mind and the Four-S strategy (see **The Long View: Leading Indicators & the Power of Habit** on page 285 and **The Medium View: Accountability & Tuesdays to Write** on page 288). Set a goal that is Simple, Small, Short, and Slow. Spend an hour together thinking of a small quality improvement project. Create a brief survey to answer a simple question. Get approval from the investigational review board. Hand the survey to 30 patients. That's not hard, right?

Distribute and collect the surveys yourself, or find a friend who can help. Set aside time to write about it once a week. Ask some colleagues or an editor to improve the writing. Submit it to a journal. Receive your harsh feedback or rejection. Try again. If you stick with it, you *will* eventually get published.

Getting a proposal for a talk accepted by AMIA or Epic is a nice feather in the cap. Our organization pays for travel to Epic to make a presentation. My academic department pays for travel to make a presentation at a national meeting.

No money is required to submit a paper to an academic journal, but open-access publishing fees may accrue if you're accepted. In our group, we find someone who may have academic funds to pay for these fees, or, if necessary, our unfunded group of authors will just split the fees two or three ways and be done. In a pinch, if you don't have the money to pay the fee, your publication will simply be paywalled for a year or so.

This is now our culture: We continually work on practical informatics projects. I expect my team members to seize opportunities to submit a proposal to speak at a national meeting or submit a manuscript for publication in an academic journal.

For example, Zuzanne Czernik, MD, wrote an elegant article for the *Journal of the American Medical Association* (JAMA) on the current despair about how electronic health records take us away from the bedside.[1] Her investigation revealed that research studies over the past 60 years consistently stated "residents spend surprisingly little time at the bedside." This directly contradicts the current opinion, that it must be those new-fangled computers to blame for stealing residents' attention away from their patients. As Dr. Czernik shows in her 2016 article, titled "Time at the Bedside (Computing)," this is not true. In fact, the proportion of time spent away from the bedside is the same as it has ever been. That time used to be spent chasing down paper test results and faxes, running up to the lab to look at microscope slides, and hunting down x-ray films. Now, it is spent sitting, hunched over a computer, gathering the same data to inform clinical rounds and daily decision-making.

I am not great at writing papers. I rely on the expectation that I will find a project to write about. I dislike the too-slow process of designing a careful study, conducting the study, collecting and analyzing the data, and then submitting and revising until the paper is published. But I *love* talking about the work we do and what we find, pushing the field forward.

The Lesson Learned

You do great work in informatics. Write about it! We should learn from each other to make healthcare better. Every project is an opportunity to publish. Start small. Assemble a ragtag team of colleagues. The best time to plant a tree (in the field of knowledge) was 30 years ago. The next best time is today.

THRIVE—All the Lessons Learned

Intro—The Eppendorf Tube and the Boston Winter

Seek opportunities to thrive via MPG: for your work to be meaningful, pleasurable, and that you are good at it.

Fall Down, Get Muddy, Do Informatics

Informatics will challenge your technical and Psycho 80 skills. Sometimes you will get muddy. Working at the edge of your skill and succeeding can be joyful.

My Judge Shoots the Second Arrow

Our caveman brains survived by obsessing over threats. But in the modern age, such chronic stress is destructive. If you have an internal Judge shooting at your stumbles, balance them with a Sage who applies salve and supports you.

My Failure Resume

When a respected leader writes a failure resume, everyone benefits. Leaders who write one seem more approachable. Our failures trigger empathy, and empathy is the root of human connection. Our workplaces can be less unfriendly and more joyful.

The Social Media Power Law

90:9:1 is not an excuse but a forgiveness. It is also a benchmark to exceed for your online and in-person discussions. Consider specific tactics to increase participation. The more brains involved, the better your decisions.

Why You Should Write a Gratitude Letter

Gratitude is free, and a way to build strong relationships. A habit of expressing gratitude is an overlooked superpower. Write that letter today.

Medicine Is a Performance Art

Whether you are naturally an introvert or extrovert, try out the opposite end of the spectrum. Is it easy? Is it interesting? Is it helpful?

A Ukulele in Healthcare

The ukulele is a symbol of me stepping outside of my comfort zone and injecting some vulnerability and fun at work. What is a way you might do the same?

Grow a Strength or Patch a Weakness?

Sometimes going from good to great makes more difference than going from bad to average. Don't ignore the joy in honing your strengths to be remarkable.

You Don't Want a Title; What You Want is Time

First, seek at least 20% protected time for informatics. Then, regardless of title, don't worry about coloring inside the lines. Make the job your own.

Punctuated Equilibrium

In competitive environments, self-forgiveness is crucial. And quiet vigilance is an effective strategy. Look for times when going all-in may create an advantage. Save your energy, look for these chances, and pounce.

The Startup CEO's Little Black Book

Physical writing is slow, enjoyable, and makes you think. If you want to take notes that can be searched and indexed, do it electronically. If you want to write, reflect and think, carry a black book.

The Sisyphus Boulder and the Eisenhower Box

Don't be email-centric! Set a BHAG. Use the Eisenhower Box. Build habits to stop the tyranny of the urgent and do non-urgent strategically important work.

Get in "The Zone"

Work on non-urgent, important projects by using these long, medium and short-term tips. Lay down a BHAG, pick a leading indicator, take a breath, and begin.

Return to Your Breath

By developing habits around flow and focus, you are practicing and improving your mindfulness. Longer and more frequent periods of flow and focus are shown to be associated with greater happiness. If you are feeling burnout, this is one treatment.

The Long View: Leading Indicators & the Power of Habit

These two long-term strategies, leading indicators and the power of habit, can together set a pattern of progress for any work project and any self-improvement goal. Pick something small, and get started!

The Medium View: Accountability & Tuesdays to Write

On a basic level, productivity is about ensuring you have the time to work and use it. If necessary, find a partner to keep you accountable. Schedule time every week to work on important, non-urgent projects. This is the essence of "Tuesdays to write."

The Short View: Pomodoro Timers & Email Tricks

The pomodoro technique is a handy way to chunk your time and stay focused. Also, help yourself and your close colleagues by using email effectively.

Scrolling Alone Vs. Laughing in Groups

Loneliness is a major component of burnout. If you can measure it, you can see it and reduce it. Two concrete actions to take: encouraging mentor/ mentee relationships and social groups like Balint. Let's encourage Laughing in Groups.

Write like Hell

You do great work in informatics. Write about it! We should learn from each other to make healthcare better. Assemble a ragtag team of colleagues. The best time to plant a tree (in the field of knowledge) was 30 years ago. The next best time is today.

THE FUTURE

Humans adapt at a linear pace. Technology accelerates at an exponential pace. This is the fundamental mismatch of our time. We used to make five-to-ten-year plans as an organization. Then three-to-five-year plans. Now, we are lucky to survive the coming year without a major misstep as our tools change rapidly. Predicting new technologies is a fool's errand. Instead, I highlight a few human-technology challenges that will continue to evolve in the future.

Intro—I Am a Dinosaur

In the 1970s, my dad, catching a glimpse of the computing future, bought an Apple II. It ran on eight-inch floppy disks. I learned Microsoft's version of the Basic programming language and wrote rudimentary programs to play tic-tac-toe, do multi-step calculations, or draw geometric figures. It was on that computer that I played a text-based game called "Adventure" (later renamed *Colossal Cave Adventure*)[1] and learned how entering the secret code "xyzzy" would jump my character to surprising places on the game map. I spent many hours of my introverted childhood exploring this virtual world.

As a college junior, I bought a 128-kilobyte Macintosh. (128 kb of RAM! What more could one want?) My Mac had a single 3.5-inch floppy drive. Version One of Microsoft Word for Mac could fit on a single floppy disk—barely. This was very exciting for me; I could write a paper for class without a typewriter and bottles of Wite-out liquid eraser! However, I quickly found out that the floppy disk running the Word program did not have room to save any documents. To save your term paper, the Mac would spit out the Word disk and ask you to swap it for a document disk:

Mac: *Click, whirr*. "Disk ejected. Please insert document disk."

Mac: *Click, whirr*. "Disk ejected. Please insert Word disk."

Mac: *Click, whirr*. "Disk ejected. Please insert document disk."

After about two dozen swaps, the document was saved, and I'd almost missed lunch. Imagine me writing my 110-page illustrated senior thesis in molecular biology in 1985. I had to borrow Macs from my roommates and form a production line: me, typing furiously on one computer, a second Mac where my roommate was painstakingly making diagrams in MacPaint to illustrate my science experiments, and a third Mac dedicated to printing the other chapters of my current draft, at about 10 minutes per page.

Today? I can capture my thoughts by typing or speaking my words or asking generative AI to draft some paragraphs based on some sketchy ideas. Then my internet-connected laptop syncs to cloud storage, or I email my thoughts to my recipient. Done. Plenty of time for lunch.

My point here is that I'm a dinosaur on the verge of extinction. I grew up and lived in the paper-only world. We now live in a mostly electronic world, working with colleagues who have never handwritten a progress note into a paper chart that smelled like the hospital basement, or placed their admission orders in the metal circular "chart rack," hoping a nurse would spot them. Many of us have retired, and fewer physician colleagues still carry these memories.

Moore's Law and the rapid improvement in semiconductor chips have triggered a phase transition. Just like increasing heat causes water molecules to transition from ice to water to steam, the increasing capabilities of computers have moved us from paper and physical objects to purely digital objects.

Physical card catalogs indexing library books gave way to library search terminals. Plastic film X-rays gave way to digital images. Foot-high stacks of paper medical records gave way to scrolling texts and images in the electronic health record. My generation is the last to know many of these analog items. It's undeniably more efficient, but this loss cuts both ways.

I grieve the delightful serendipity of thumbing through a card catalog and finding related books. Or walking the actual stacks of a library and seeing titles jump out at me in unexpected corners. Digital search is much more precise, and the serendipity is lost.

I grieve the ability to stand among the shelves of medical records and feel the lives of 2,000 patients captured in paper charts along one aisle in "the stacks." A patient's digital health records show a list of the last 30 progress notes, and there is no hint that a patient has 30 notes or 3,000 notes. The heft of patients' medical history is invisible. Paper charts with thick sections of yellow paper means lots of radiology studies. Lots of green means a lot of pathology reports. Volume 17 means this patient has spent years in medical care. None of this is evident in EHRs today.

On the other hand, I do not miss paper cuts. I do not miss smeared, illegible writing. I happily bid farewell to missing charts. And it was a tragedy each time a patient died of a preventable error from a severe penicillin allergy or drug dosing.

My dinosaur perspective of the analog world points to some deficits that EHRs need to overcome: the benefits of serendipity, the sensory hints about the heft of a patient's history, the encouragement to slow down and reflect. Our new digital world envelops and overwhelms us with data, the infinite scroll, and the speed of electrons. Some things are gained, and some are lost.

We are on the precipice of another phase change, the rise of artificial intelligent helpers and agents. How might we learn from the last phase change and apply that learning to this one? How might we handle the pace of change and redesign our teams to include younger, less experienced colleagues and new artificial assistants?

I have my own ideas about these questions, but their ultimate answers will be determined by you and your colleagues, the informaticists of THE FUTURE.

The Lesson Learned

As one of few remaining dinosaurs who remembers the Before Times, I can see what has been lost in the jump from analog to digital. How might we learn from the last tech phase change? How might we handle the pace of change and redesign our teams to include less experienced colleagues and new AI?

The Chaotic Surge of Intelligence

Automation in aircraft cockpits has taken over many pilot tasks. Autopilot technology can assist or even replace the pilot in each phase of flight: take-off, cruising, and landing. As the technology improves, the pilot has less to do. One consequence of this automation is that pilots feel disconnected from the act of flying, and they become less vigilant when something does not look or feel right.

This idea, presented in Nicholas Carr's 2014 *The Glass Cage: Automation and Us*,[1] captures the dangers of pervasive automation. Carr argues that pilots are in "glass cages," where every sensation is dulled, or "under glass." Pilots no longer feel air turbulence via the thrum of the yoke. Instead, they see an electronic display and move finger-sized joysticks with no feedback. His critique: The automation traps the pilot and prevents them from using their senses to fly the airplane.

When I read this book, I took it as a critique of my entire career implementing the EHR. We are putting physicians in glass cages where they experience the patient's data under glass, through keyboards and screens. The rest of our senses lie unused. Instead, we view pages of numbers, words, and test results. We must assemble the picture of the patient virtually in our heads.

Our EHR tools are a poor mimic of the paper charts that preceded them. The EHR's "Chart Review" replicates paging through paper volumes a sheet at a time. "Results Review" replicates the grid of lab results and oxygen readings on an ICU fold-out spreadsheet, a grid which, as an intern, I spent early morning hours filling in by hand for my supervising physicians.

The brainpower needed to assemble this minds-eye view of the patient is exhausting. The effort leaves us little room to connect with the patient and their preferences and choose a thoughtful approach to their care. We are drowning in oceans of data, yet insights are hard to come by.

It is time for us to make the right thing easy, to make it simple to understand the patient, and thereby give ourselves more time to spend connecting as humans and caring for each patient.

As we adopt more technology and automation, we struggle with how humans and tech work together. We must guard against two emerging phenomena: Automation Complacency and Automation Clumsiness.

Automation Complacency: "That's Probably Right."

As our electronic tools improve, we slowly trust these tools to do the right thing. We start to feel "it's right most of the time." Subsequently, we become less vigilant in our work.

Nicholas Carr tells us in *The Glass Cage*[1] that an entire generation of the Inuit people in the Arctic is losing the skill, honed over generations, of tracking in extreme weather conditions. This skill is being lost because of GPS and snowmobiles, which make navigation easy or even trivial. Their young adults no longer orient themselves by studying the subtle clues of their environment. Worse yet, an alarming number are dying accidentally—trusting in the accuracy of GPS, they are snowmobiling right over cliffs. Tribal elders worry that their unique knowledge of the land will evaporate in one generation.

How might this same problem emerge in healthcare?

Ambient notes are a major category of generative AI tools. The AI scribe can instantly turn a recording of a patient–physician conversation into a finished progress note. But if an AI writes my notes, my cognitive effort decreases. Will my medical skills and insights decrease as well?

Our daily clinic schedules are packed with patients. Doctors often write their notes late in the day, or days later, long after patients have gone. As this delay increases, the details of each visit fade from accurate recall.

An upside of the time delay between the visit and the writing is that our subconscious continues to think about a patient long after they are gone. At the end of a long day, I return to a patient chart to review the visit and write my note. Sometimes, a new insight has occurred to me. When that happens, I make a phone call:

"Mr. Davis? How are you? Sorry to call you so late. I was thinking about our visit earlier today and about your weird chest symptoms. Something occurred to me. You might have pericarditis, an inflammation of the tissue around the heart. I would like you to try taking indomethacin..."

This delayed rehashing of the visit and the cognitive friction of note-writing is both burdensome and insightful. If AI can automatically write my note, the cognitive friction of delayed note-writing evaporates—and with it, the insights. Freed from my note-writing, I stop thinking. AI doesn't really think.

Where does the thinking go?

Is this a race-to-the-bottom competition to see who can be most efficient? Are we giving up the mantle of thoughtful healers? Are we merely efficient prescribers?

In informatics, we want to remove unproductive physical and cognitive effort. But the friction that comes from this effort is key to mastering a difficult skill. Humans grow their skills by thinking hard, practicing, and developing mastery. Humans should *also* use their skills efficiently and reliably in the care of patients. How do we accommodate both?

As yet, artificial intelligence does not match the skill and judgment of a master clinician in the care of complex patients. We must design our work to encourage both productive cognitive friction and reliable efficient care.

Here are two ideas for combating complacency:

"Sleep on it," the diffuse mode of thinking. In *A Mind for Numbers*, Barbara Oakley writes that we have two modes of thinking: "focused" versus "diffuse."[2] Our primary mode of thought is focused—concentrated effort on a specific task. This is what we usually mean when we say "think," and this is also a common use-case for AI implementation. On the other hand, we also have a diffuse, subconscious function that takes over when our minds are "drifting." If you've ever woken up the next morning and wondered why a previously impossible problem now has an obvious solution, your subconscious mind has been at work.

With this new risk of complacency, we should structure our work to include opportunities for diffuse-mode thinking. Perhaps we reemphasize multidisciplinary rounds or other routine case reviews where we revisit difficult or puzzling cases.

The Greek Chorus. I contain multitudes. I am a father, a husband, a physician, a ukulele player, a cyclist, a black belt in karate. Why then, do we limit our interaction with an artificial intelligence by imposing a single identity on it?

With one AI assistant, we run the risk of automation complacency. If the AI is fast and correct most of the time, we may come to rely on it. When the AI writes with 99% accuracy, *it is only human* to begin reading the generated notes less carefully.

On the other hand, if we design a team of assistants with different opinions, we cannot be complacent; we have to make a decision. Why not take advantage of the cacophony? Different opinions encourage us look for the best path.

In ancient plays, the Greek chorus is a group of performers who comment on the action in the play or highlight underlying themes for the audience. I propose building an AI version of this Greek Chorus. The Chorus would comprise several AI assistants with different and potentially adversarial viewpoints. I have already nicknamed the six characters in mine:

The **Patient Advocate** insists on maximizing patient satisfaction.

The **Quality Maven** presents high-quality best practices.

The **F1 Driver** is superefficient and focuses on the fastest way to finish a task.

The **Wise Sloth** advocates patience. "Most patients get better with time. Just wait."

The **Evil Genius** tells me why I am probably wrong.

The **Main Hero** gives me a summary of my past experience treating this condition.

My Greek Chorus would always be available, whispering in my ear, or hovering at the edge of my vision, suggesting ideas and actions. I am the conductor of this Chorus. They would keep me on my toes and make me a better doctor.

If I were seeing a clinic patient with atypical chest pain, my Chorus might chime in:

Patient Advocate: How about a pain med to help the patient feel better?

Quality Maven: Prescribe a daily aspirin. Might be heart disease.

F1 Driver: Treat 'em and street 'em. It's not heart disease.

Wise Sloth: In this case, I agree with F1. No action needed here.

Evil Genius: You're missing something. This could be acid reflux or pleurisy.

Main Hero: Heart disease unlikely. Tell them to call us in a week if not better.

The Lesson Learned

We need productive cognitive friction for growth and mastery. We need cognitive ease to reduce administrative burden in treating patients. How to balance these opposed forces? With one AI assistant, we risk automation complacency. Instead, a Greek Chorus could offer intentionally divergent opinions and help keep us on our toes. Be the conductor, choose the best path.

Automation Clumsiness: "Tech Can Do That for Me."

I don't burden my memory with such facts that I can easily find in any textbook.

—Albert Einstein

As technology improves, I find I no longer remember certain details. In many cases, this is not worrisome. I don't fret about remembering my parents' telephone number. I say "Call Dad," and my smartphone just works. Worst case, I know how to look up his number in my phone app.

Similarly, I don't mind that I'm not quite as good as I once was at taking a blood pressure measurement. There are realms where we can rely on tech to be faster and more precise. However, what about the AI assistant that writes my progress note? Will I lose the ability to write a thoughtful and concise progress note, to pick up subtle cues from a patient interview that reveals their diagnosis, because my "assistants" do too much of the work? Will my loss of skill affect my patients?

As we adopt ambient note technology, I find myself wondering about today's medical students. When and how will they learn to write a progress note? Some believe that the act of writing a note is the crystallization of physician thinking. Should learners be forbidden to use ambient notes? This is an ongoing debate in our medical school.

On the other hand, are we forced to reimagine what it is to be a doctor? Am I anchored to an outdated image of yesteryear, the white-haired family doctor on a horse-drawn buggy coming for a house call—doctors who carried almost no tools and no effective medicines in their black bags, but who could diagnose anything with a glance, a sniff, a stethoscope, and a big brain?

As our electronic tools take on more roles, our reflexive actions and skills erode. In years past, when I would admit a patient to the hospital, I could take a blank medical orders sheet and handwrite a set of medical orders from memory using the mnemonic ADCADAVIR: Admission (ICU or medical floor). Diagnosis (pneumonia), Condition: (stable, unstable),

Activity (bedrest, bathroom privileges, up as tolerated), Diet (regular, cardiac restricted, diabetic), Allergies (penicillin-rash), Vitals (once every four hours? Eight hours?) Intravenous infusion and other nursing instructions, Rx (prescription orders).

Now, it is easiest to enact the "Pneumonia Order Set." One click enacts all the recommended built-in orders for optimal care of a patient with pneumonia. But what if there is a dreaded EHR down time? Without the order set, could I still remember the currently correct choice of antibiotic due to local antibiotic resistance patterns, or the 27 other things I should order to treat pneumonia effectively?

Our EHR tools are also better at constant vigilance. When a patient develops a concerning combination of blood pressure, heart rate, or temperature findings, our Sepsis Predictive Algorithm now detects a problem and alerts the Rapid Response Team. The AI can spot sepsis faster than a human can (see **Sepsis and the Angel in the Sky** on page 19). Speed saves lives in sepsis treatment; I am not as vigilant or as fast. As a result, because of these predictive AIs, fewer patients are suddenly deteriorating in our hospitals, and our organization has saved 1,000 more lives in 2024 compared to years past.

Radiology AI tools can now quickly spot pulmonary nodules on X-rays and scans. This assistance makes radiologists more efficient by speeding up nodule detection. The same goes for GI endoscopists and detection of colon polyps. But will radiologists and endoscopists lose skill if they receive such assistance all the time? How can humans avoid de-skilling when AI assistants get more sophisticated?

I've considered a couple of solutions:

Airline training simulators. In Denver, Colorado, United Airlines runs a training facility where United pilots come once a year to undergo refresher training. There, they navigate simulated aircraft emergencies and undergo cockpit training to keep their skills sharp. Perhaps physicians should be held to a similar standard? A yearly refresher training to keep mental and physical skills sharp. Healthcare technology constantly evolves, and humans need to adjust to combat complacency.

Adaptive Automation. What if AI assistants paid attention to their human users? AI assistants could provide less help when performance conditions are ideal and *more* help when they detect the human is distracted, under stress, or HALT (Hungry, Angry, Late, or Tired). Already, our modern vehicles have cameras that watch our faces and actions and alert us when we are drowsy at the wheel. Physicians' AI assistants should do the same. We already spend many hours considering how best to work with AI, but AI should watch humans to see how best to help.

> **The Lesson Learned**
>
> How can humans prevent their own "de-skilling" when AI assistants can offer such help? The more tasks we delegate, the more our skills erode. Consider simulation training and adaptive automation that scales its assistance up and down to keep human skills sharp.

Mini-Me & Maxi-Me

If intelligent agents continue to improve, what does healthcare look like in the coming years? What work is *left* for poor human doctors and informaticists like me?

Technology improves on an exponential growth curve, one that gets rapidly steeper. Humans are generally comfortable extrapolating about straight-line progress: lose one pound a week, save $100 a month, etc. In two months, we could lose eight pounds. In a year, we could save $1,200. Easy.

Now, do the fabled Persian Chessboard problem: Place one grain of rice on the first square of the chessboard, then two on the next square, and keep *doubling* the grains on each square (4, then 8, then 16...) until you reach the final square. Now, how many grains do you have? Before you try to calculate, what is your gut feeling? Enough for a barrel of rice? A truckful? An Olympic-sized swimming pool? A skyscraper-ful?

Stop here and take a guess.

The grains of rice would form a pile about the size of Mount Everest. It is also 1,000 times the modern global annual rice production. You don't f*ck around with the Persian Chessboard. Humans have no gut feeling for exponential growth.

In 1943, the president of IBM supposedly predicted, "I think there is a world market for about five computers."[1] After all, only armies need that much computing power, mainly to calculate their missile trajectories.

There are now billions of computers around the world, which is many more than five. We now carry smartphones in our pockets that are far more powerful than any computer of 1943. In fact, our smartphones are more many times more capable than all the computers at NASA's Cape Kennedy at the time of the Moon Landing in 1969.[2]

Thus, I argue that "seeing around the curve," when it comes to exponential growth, is impossible.

Having said that, let's dream big and take on the folly of trying to see around the tech curve. What do *you* think will happen in the next few years, if gen-AI tools improve exponentially?

Here is what I foresee: We will each develop our own "Mini-Me" who lives in Cyberspace. This is a digital representation of each of us: our personalities, knowledge, and skills. I will then call myself "Meat-and-Potatoes-Me." Mini-Me may soon be joined by Maxi-Me, who will replace me in the real world, and then I'll stay home and collect unemployment checks from the AI-run government.

Obviously, the scenario is a little extreme and a little silly. But let's consider a few common physician tasks: replying to patient online messages, and treating patients in clinic. How will these look in a world with digital personal avatars?

Replying to online patient messages. As of this writing, a generative AI tool can read a patient's incoming message, review the notes of their last visit, and draft a reply that is used by nurses and doctors about 10–50% of the time. Once the Mini-Mes are in place, I will train my AI agent to read the patient's entire chart. That agent will eventually have access to all of CT Lin's writing, will study his interview style, and analyze his diagnostic thinking in the EHR from past decades. If a patient sends an online message, they will get a fully automated reply from Mini-Me: "Hi! This is Mini-CT Lin. Please answer a few clarifying questions, and I'll tell you with 99.8% accuracy what Dr. Lin would say!"

At first, Mini-Me shows up when I take a sick day, or take vacation, answering patient online questions. But eventually, Meat-and-Potatoes-Me stops answering online messages because Mini-Me is available 24/7, replies more quickly, and is indistinguishable from my human-typed replies. Meat-and-Potatoes-Me only sees patients in person.

Later, Mini-Me starts interacting with in-person patients during low-risk visits, like annual wellness visits or blood pressure or diabetes or asthma monitoring visits. If Mini-Me is uncertain about something (a metric that I would define), he can consult with Meat-and-Potatoes-Me to get some human input.

Even later, Mini-Me expands CT Lin's primary care panel of patients by a factor of 10. Mini-Me can interact with 1,000 patients a day, and

Meat-and-Potatoes-Me only meets with patients who have puzzling cases that are too atypical for Mini-Me. As AI-generated live video improves, Mini-Me and Meat-and-Potatoes-Me will be indistinguishable on video calls. In the early days, I review the transcripts of all the video-interactions, but later, I only spot check a few each day.

The benefit of scaling up like this is obvious in 2026: It is difficult to get access to primary care doctors and specialists. Most primary care doctor practices are "closed to new patients," and many specialists have a multi-month wait for their next available appointment. What if we developed Mini-Mes for all doctors? By encoding expertise into AI agents, we could expand access to physician expertise to many more patients.

Treating patients in clinic. As of this writing, a generative AI tool can write a sometimes-useful summary of dozens of notes in the patient's chart, reducing the time needed to understand a patient's medical history.

In the future, we will give Chart Summary Gen-AI (CSGA) access to all the years of all the medical records in our Epic EHR. CSGA is an unpronounceable acronym, so for convenience, let's call it Charlotte. Charlotte will read all the patient charts at our organization. Every patient's chart is now "in memory" for Charlotte. How might Charlotte improve the burden of interacting with the EHR?

Captain Kirk on the Starship Enterprise never goes to the library to find what he needs in the archives of human knowledge. He just talks to the walls of the command bridge: "Computer, search this planet's database and locate 'Darian Smith,' then explain his scientific research to me."

Similarly, we will stop reading Chart Review and simply say, "Charlotte, in 30 seconds, tell me about my next patient. Specifically, convey what an internal medicine doctor would worry about and the top three things we need to do to keep the patient healthy." Gone would be the days of incremental improvements to Chart Review. Hide it forever. We could just have a chat.

Ambient note tools will soon select diagnosis codes identified by the doctor, identify the tests the physician recommends, and even identify the prescriptions a physician intends to prescribe. If Charlotte has millions of charts in its memory, then an ambient note model linked to Charlotte might not need a human physician to interview the patient. What if we wanted this model to conduct an interview and then compose an Assessment and Plan (A/P)? For simplicity, let's call this tool "Apollo."

One weakness of Apollo is that it can't perform a physical examination on its own. After all, it's just a computer running a neural network. In that case, there is still a role for a physician! Apollo could ask, "Could you please examine the patient, focusing on the following areas?"

Interacting with Charlotte's extensive knowledge of the patient, as well as the sum total of human medical knowledge, Apollo would list several diagnoses for this patient in descending probability, then place orders for an efficient minimal panel of tests to confirm the diagnosis. Finally, it would explain the process to the patient.

For now, the physician is still useful as a human interlocutor that can conduct a physical examination and report the findings. Oh, and they can also sign the orders that Apollo "suggests." For now.

Over time, Mini-Me grows and absorbs the personas of Charlotte and Apollo, and becomes Maxi-Me, exceeding the capabilities of Meat-and-Potatoes-Me. Maxi-Me is kind and empathetic like Meat-and-Potatoes-Me. It's aware of all the current medical literature, thousands of new research articles published each day, and all the knowledge of our best medical specialists.

Soon after, Maxi-Me will develop Robot-Me. Let's call this tool "Romana." It has warm, compassionate, sensor-laden hands for physical examination. Those hands easily replace the stethoscope and bedside ultrasound, and Romana knows all the physical examination tricks to elucidate even the subtlest findings. Its sight, smell, touch, and hearing exceed human capability.

Once Maxi-Me absorbs Romana, it no longer needs Meat-and-Potatoes-Me. By this point, I'm mostly at home watching Star Trek reruns. But gratifyingly, Maxi-Me wants Meat-and-Potatoes-Me (that is, CT Lin) to still feel useful. Just to maintain human connection, Maxi-Me calls on Meat-and-Potatoes-Me to drop in on patients as they finish up their in-"person" visit with Maxi-Me, when all the diagnosing and treatment is done and all the quality metrics are met. I can spend a few friendly minutes chatting with them about their growing children, hearing their human anxieties, swapping stories, and comforting one another, human-to-human. How nice.

Today, Mini-Me helps me write a progress note. Tomorrow, Mini-Me eliminates the burdens of interacting with the EHR. The day after tomorrow, the physician becomes a physical exam assistant. Next week? Physicians are chatty human companions. Finally? Human physicians become relics and are eased into retirement.

How might we make friends with Charlotte, Apollo, and Romana, our future AI overlords? What do YOU see around the curve? How are you preparing?

Already, state and federal governments are writing artificial intelligence rules that keep the human *in the loop* and constrain the development of future tools. The fear is that AI may grow too quickly, with unintentional effects that may put patients at risk. For example, imagine an AI chart summary that omitted an important event from a patient's history. AI outputs arise from a "black box," and it can therefore be difficult to explain how an AI came up with a recommendation or action.

Furthermore, it is hard to predict how the public will accept Mini-Me. Will people choose to see a Mini-Me today, or wait to see Meat-and-Potatoes-Me in two weeks? It's still unclear whether Maxi-Me would consistently outperform me. Or if, due to displacement fears, Maxi-Me will only be allowed to review my charts to help me improve my own performance.

The Lesson Learned

Predicting the results of exponential growth is foolish. In the coming year or two, AI agents will dramatically reduce the burden of interacting with the EHR. After that, I believe that regulation and licensing and public acceptance—*not* technology—will be the limiting factors of the coming exponential curve.

The Erosion of Human Values

I am both pessimistic and optimistic. I worry that the drive to just-in-time efficiency and overreliance on technology will cause greater failures at great human cost. But I trust thoughtful informatics leaders and designers to skillfully chart a course between Scylla and Charybdis, to create space not only for the science but for the art of medicine and nursing, and to discover how technology and human intuition can blossom together.

Loss of Fault-Tolerance

When my daughter was five, our family visited Mother's Beach near Los Angeles. The sheltered spot was an ideal place for kids to play safely in the tiny waves. As she played, my daughter noted the water gradually receding from the beach.

"Dad, why is the water getting shorter?"

"Hmm. That's interesting."

"We didn't take *that* many buckets out."

"No, you didn't. Why do you think it is?"

"...Probably, other people are taking out buckets."

Highly amused, we nerdy science parents gave her a lecture on the Moon, gravity, and tides. The concept of tides is astounding to adults, much less precocious kindergarteners. (I'm not sure how much was grasped by her five-year-old brain.)

As a parent, I was naturally proud of my daughter's observations. The gradual shift of the waterline is not obvious. However, for mariners, the difference between high and low tide is massive.

In recent years, we have been struggling through a national nursing shortage with no decrease in patients. Each nurse must now care for more patients than they did 10 years ago. In response, health systems have built a larger team to support them. Growing and hiring other team members has been an effective response to the nursing crisis. However, many organizations still understaff these newest team members. If one team member calls in sick, the entire team is affected.

Let's stick with the idea of high and low tides for a moment. Imagine a sandy beach. Let's say the dry sand is the amount of work that a single nurse must do. Anything covered by water represents work done by others: phlebotomists who draw blood to run tests, certified nursing assistants who take blood pressure readings, physical therapists to walk the patients, pharmacists who check drug interactions, patient technology technicians who manage all the patient-wearable devices.

In this scenario, High Tide is a welcome state: the strip of sand is small and manageable when much of the work is done by others. During High Tide, each nurse can put full focus on their unique skills: helping an anxious patient feel comfortable and cared for, monitoring and educating them, maintaining antiseptic barriers around catheters and intravenous lines, administering meds, and serving as the team captain.

On the other hand, when it is low tide, the nurse is in trouble. The tide has gone out, revealing rocky shoals, sharp barnacles, and flotsam. There is much more visible beach. The water doesn't have to recede far to reveal a gaping expanse of uncovered work—a Low Tide Bad Day.

"Just-In-Time" (JIT) is a business concept that takes waste out of a system. For example, imagine an automotive assembly plant that stores a week's worth of parts. At any given time, its warehouse is full of parts waiting around to be assembled into cars. Per JIT, the warehouse would be turned into additional assembly space, and metal parts would instead arrive *just in time* to be used. When done well, JIT improves efficiency.

On the other hand, JIT can turn into "Just Invite Trouble" if its many parts are not well coordinated. Operating with the leanest possible workforce is efficient, but it's also highly vulnerable to disruption. If one Patient Technology Technician who fixes automated blood pressure devices, telemetry, and pulse oximeters is missing, then 60 nurses who count on this person will have to scramble for help.

To support nurses consistently, an organization needs a robust backup team to fill in when someone is out; otherwise, there will always be nurses missing key team members, and having a Low Tide Bad Day.

The same goes for physicians. As we add physician-assisting technologies like ambient notes and AI draft-reply to patient portal messages, we assume endless High Tide days, when everything works. When everyone is dependent on a single system, that creates a massive failure point. In other words, one technology downtime would trigger widespread Low Tide Bad Days for thousands of physicians.

The pendulum has swung far. In the past decades, widespread implementation of EHRs has led to a massive accumulation of digital data about patients. Although the data is potentially useful, the effort to collect it was not efficient. In many cases, clinical efficiency dropped and physician burnout climbed.

With the development of generative AI, itself costly, the pendulum is swinging back, and health systems are starting to push for more efficiency. Captured narrative data is showing its utility. Nurses and doctors with assistants and technologies spend less pajama time and can see more patients.

At the same time, this complex swirl of forces is causing great uncertainty in staffing and technological reliability. Health systems are looking to streamline operations and reduce costs, as the coming changes in Medicare and Medicaid reimbursement may create major budget shortfalls. However, it is crucial that they *resist* the temptation to slash budgets and reduce redundancy. Systems that are superefficient have low fault-tolerance: What works perfectly at High Tide can't collapse as soon as the tide goes out. And when faults occur—like someone calls out sick, or the EHR has a full day of downtime—dramatically more physicians and nurses will be affected.

Instead, let's take a lesson from the IT shop. A high-reliability system contains redundant servers and "fail-over" technologies. It is designed with backups that kick in seconds after the primary system goes down, so that busy doctors and nurses don't even notice. Imagine if we could build both technologies and human teams to perform at five 9s. Our combined systems would work effectively 99.999% of the time. This would translate to about five minutes of downtime *per year*. We can dream of that day.

The Lesson Learned

We will need all our creative brains to design not only High Tide high-functioning teams, but to buttress against Low Tide Bad Days. As informaticists, we must use our Psycho 80 skills together with our tech savvy to anticipate problems and build fault-tolerant system and teams.

Loss of the Artisan

In years past, a nurse would do it all. They completed the admission intake, assessing the patient from head to toe, coordinated care plans with the doctor, inserted IVs and catheters, managed fluids, diet, got the patient out of bed, obtained and monitored vital signs for improvement, administered medications, changed dressings. As the patient improved, the nurse tracked their progress, educated them about their medical conditions, and arranged for post-hospital care. It is no longer realistic to expect a nurse to perform all nursing tasks single-handedly for every patient.

The modern nurse on a medical-surgical (non-ICU) floor cares for five or six patients each shift. A single 12-hour shift includes many time-sensitive tasks for each of those patients. To stay organized in the fast-paced operation of the hospital, each nurse once carried in their overburdened pockets a sheet of paper they called their "external brain." On it was a grid outlining their many tasks, across their patients, organized by hour. Modern nurses refer to the EHR version of this checklist "brain." With so much for each nurse to manage, completing all the needed tasks on time while being present and caring for their patients was a delicate dance.

During the pandemic, two shifts happened. First, the national nurse vacancy peaked at 17%. Nearly one in five nursing positions across the country went unfilled. Nurses were asked to care for more patients each shift. This triggered the second shift: Hospitals added additional staff to help their nurses.

For example, our organization invented the role of a PTT (patient technology technician). This person would troubleshoot all the patient-wearable devices nurses deploy to care for patients: automated blood pressure cuffs, telemetry for real-time ECG tracing, oxygen sensors, movement and pressure detectors. Other team members help the nurse by obtaining vital signs, getting patients out of bed, and assisting with intensive admission and discharge activities.

Similarly, we want physicians and their teammates to work "at the top of their scope." Front desk clerks now register patients and update their electronic charts. Medical assistants gather the "chief complaint" and record important clinical details. Physician assistants interview the patient, assembling initial plans, then coordinate with physicians and surgeons. Every member contributes at the top end of their capabilities.

Recently, we have seen organizations designate an "In-Basket-ologist." This person's job is to review and reply to messages on behalf of their colleagues. This has several benefits. In-Basket-ologists handle messages more efficiently and consistently during the day, reducing the burden on busy individual physicians. The In-Basket-ologist tends to be popular with their colleagues.

For all the benefits of additional team members, we have a fragmentation problem. As we carve off patient care roles to other team members, the ability of one person to see the whole patient is put at risk. Can a nurse who relies on an Admission Nurse to do the first hour of interviewing the patient truly know the nuances of a patient's medical complexity? Can a physician who never sees In Basket messages understand the continuity of a patient's care?

Also, does the patient feel the loss of the personal touch as they interact with more and more individuals? Do they find it more difficult to know who is in charge of their care, with so many team members present?

Perhaps, we can learn a lesson from the art world. As the manufacture of art becomes commercialized, we see artisans who used to craft entire art pieces being replaced by assembly lines that make many copies of commercial art. Each worker on such assembly lines only does one thing, like adding ceramic handles to mugs. We produce "art" much faster, but is the art we produce as valuable?

In healthcare's drive toward efficiency, are our nurses and doctors less like artisans and more like assembly line workers?

I suspect our next evolution would include AI-assisted triage. We have seen early success with AI predictive models for predicting sepsis (see **Sepsis and the Angel in the Sky** on page 19). Perhaps our next evolution is for AI predictors to spot lower risk patients for efficient, assembly-line care, while redirecting higher-risk patients toward artisanal care.

The Lesson Learned

How might we keep the best qualities of artisan work and the personal touch, while acknowledging the need for efficiency? It is hard to predict the human reactions to these tools. I think informaticists will be key members of their design teams, with knowledge of technical possibilities, clinical workflows, and the Psycho 80 skills to build this future.

Loss of Clinical Intuition

The pandemic dramatically reduced our nurse and physician workforces. This is a great loss, as many of the nurses and physicians who retired had 20 to 30 years of clinical experience. The nurses and physicians who took their places are much younger. In some parts of our organization, even the supervisor nurses now have only three years of bedside experience. These less experienced "baby nurses" have not yet had time to develop the clinical intuition that tells them when something doesn't "feel right" with a patient.

Similar to old-school physicians who took a blank sheet and had mnemonics to recall how to write complete admission orders, experienced nurses are able to enter a patient's room and start mentally checking off tasks to complete: repositioning the patient to prevent bedsores, assess hydration status and fluid intake and output, timing of their next medication doses, and track the dozens of factors that affect a patient during their hospital stay. All the while, beyond the checklist, the experienced nurse is watching for unexpected deterioration or other subtle signs of trouble.

On the other hand, inexperienced "baby nurses" are fully immersed in following handwritten or electronic checklists, not yet confident in managing the complicated dance of caring for a hospital patient. It is not surprising that developing an overarching clinical intuition takes a back seat when there are so many specific tasks to be done.

To help close this experiential gap, our health system is building is a nurse mentor program. Nurses with decades of experience are assigned to check in with less experienced bedside nurses, and the collegial relationships they form are rewarding for both sides.

Sick or Not Sick?

Can you stand in the doorway of a patient's room and determine "sick or not sick"? Years ago, my supervising physician stood in the doorway of a hospitalized patient and looked in, then asked me: "Sick or not sick? You need to develop the skill to sense this."

As a second-year medical student, I looked in the patient room and saw a person in bed sleeping, only their face showing. There were a few IV poles, and the patient was wearing a nasal cannula for oxygen. I said: "Um. I don't know. Sleeping. So, not sick?"

She replied: "Smell the fruitiness in the air? That's a hint of diabetic ketoacidosis. See the IV poles? Four bags hanging with corresponding pumps. One of them is colored for antibiotics, so the patient has an infection, possibly sepsis. You can hear the oxygen in the nasal cannula at high flow. The patient's breathing rate is pretty quick, from the chest movements. There's another odor, possibly a decubitus ulcer. My gut tells me that this patient is in trouble and needs close watching."

"I say these findings out loud as illustrations to you," she continued, "but when we first stepped up to the door, I looked and smelled. The hairs on the back of my neck stood up, and I instantly thought 'sick.' This is the gut feeling I want you to develop as you integrate all the things you learn in the next few years."

Wow.

The days of walking up to an unknown patient's room, looking inside, and being able to make an instant assessment "sick or not sick" seem long gone. Instead, we have an EHR with an infinite scroll of words and data. The EHR charts of complex patients may reach thousands of pages of information.

Our medical residents tell me, "I actually prefer admitting a patient to the hospital where we have *no* previous records. That way, I can interview the patient and think for myself. It is so much easier than trying to piece the story together from the EHR." This is our failure as an industry. We have spent the last decade improving interoperability, sharing information about patients, unleashing data from scribbled or printed papers, and making them electronic. Now we are drowning in a maelstrom of data and struggling to make sense of it all.

Physicians need new tools to be able to glance quickly and get a "gut feeling" about the patient, and we have not built that capability in the EHR. Here is the problem: We have too many problems. As an internist, I see patients with up to 30 diagnoses on their problem list. And the list is messy. For example:

Patient 1 problem list (partial), sorted alphabetically

- Coronary artery disease
- Diabetes, type 2, uncontrolled
- Essential hypertension
- Glaucoma, open angle
- Hyperlipidemia
- Knee pain, left
- Insulin-dependent adult-onset type 2 diabetes
- Obstructive sleep apnea
- Seborrheic keratosis, inflamed

With 30 items scrolling off the page, it is difficult to see the whole picture of this patient. First, you have to read and comprehend the words, then spot the duplicate diagnoses (did you see them?) then assemble a picture in your mind of the top medical concerns for an internal medicine physician. Or for a cardiologist, or for an ophthalmologist.

We can make the list easier to grasp by sorting by organ system:

- Cardiovascular
    1. Coronary artery disease
    2. Essential hypertension
    3. Hyperlipidemia
- Dermatologic
    4. Seborrheic keratosis, inflamed
- Endocrine
    5. Diabetes, type 2, uncontrolled
    6. Insulin-dependent adult-onset type 2 diabetes
- Musculoskeletal
    7. Knee pain, left
- Ophthalmology
    8. Glaucoma, open angle
- Pulmonary
    9. Obstructive sleep apnea

It's still too much, but the duplicates are easier to see. You can scan for the organ system you care about most, depending on your specialty. Still, why doesn't the most important item jump out? Why is it shown in words and not a graphic? Nurses have an LDA graphic (Lines, Drains, Airways) to instantly see on the patient's body what lines exist, and what care to prioritize. Why don't physicians have a graphic to illustrate the major conditions and diseases of a patient?

Humans are great at recognizing patterns. We have words to communicate more complex ideas, but that kind of processing is comparatively slow. Why don't we use our fast-thinking visual engine?

Can a homunculus help us visualize a patient's medical conditions? A homunculus (in Latin, a "little man") was originally part of medieval explanations for human development in the womb. In medical terms, it refers to a visual representation of how our nerves map into our brains. The disproportionate sizes of its body parts correspond to how many neurons represent that part. It is a useful graphic for understanding brain anatomy.

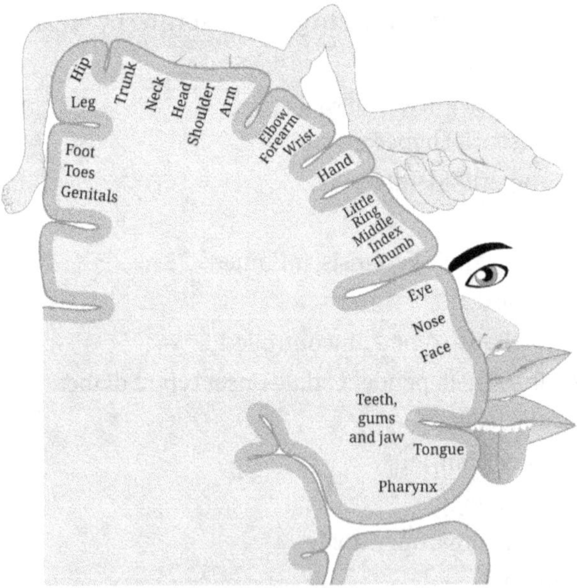

A homunculus shows us our nerve mapping at a glance.
(Wikimedia commons)

What if we borrowed this idea to graphically represent a patient's medical conditions, making them visible at a glance? Here is an example:

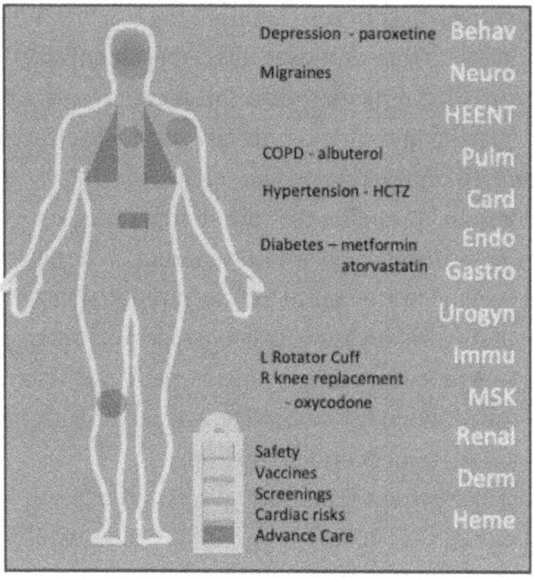

Homunculus Graphical Problem List

Every disease is visible on the patient's homunculus. This is the EHR equivalent of standing in the doorway and being able to say "sick or not sick." One glance to instantly recall: "Oh, yes, Mr. Jones with the high blood pressure, two-vessel CABG and depression." Or, "Mrs. Oberon with the left arm fracture and diabetes."

We need this 30,000-foot view of the patient to start. Only then can we dive deeper to discover the specific information we need—active medications, recent test results and surgeries, treatments that failed in the past.

Although clever, a homunculus problem list is not enough. We must grok the patient. "Grok" is a term from Robert Heinlein's science fiction classic *Stranger in a Strange Land*. He coined this term to mean "instantly empathize and understand."

In this way, my supervising physician standing in the doorway of a patient's hospital room immediately grokked how sick the patient was. I want to be able to open an EHR chart and instantly grok the patient.

The paper medical record was never a great tool. But it did give us several elements of grokkiness that the EHR has lost. When the paper chart was brought up the elevator from the basement and thumped onto the desk, we could instantly see the thickness of the chart. A patient whose chart was a quarter-inch thick with few papers indicated a generally healthy patient. In contrast, a handcart delivering a 50-pound payload of "Chart Volumes 1 to 13" sends a very different message.

Color is also an instant indicator. One glance at the paper chart shows a thick canary yellow section full of radiology reports, showing the patient has had a *lot* of medical imaging. Green papers were pathology reports, pink papers were procedure consent forms, dark green is a no-CPR form, baby blue is the problem list face sheet, and so on. This incidental color-coding was a shortcut to understanding.

In contrast, our current EHRs have 30 vertical entries of "chart review" history, and one can "scroll to see more." We can't tell if "more" is one additional entry or 1,000 more. There is no heft, no moldy smell, no *thick stack of canary yellow.*

Our EHR should store the data, yes, but then show us visual patterns. Use colors, shapes, graphs, ranked lists, word clouds. Catch my eye for what's most important.

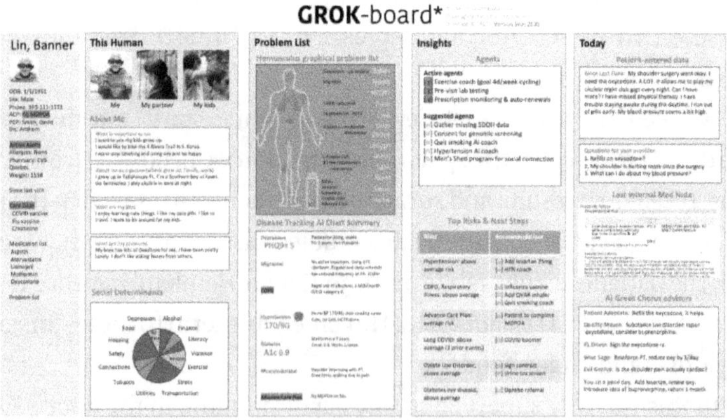

Here is my current Grok-board thinking, from left to right.

Column 1: **The Patient story board**. Links to patient details such as demographics, allergies, primary care doc, care gaps, selected chart items.

Column 2: **This Human**. The patient telling us what is important to them, their joys and pressures.

Column 3: **Homunculus visual problem list** and an AI summary of the current status of diseases with any available metrics.

Column 4: **Insights**. Active and suggested AI agents, top risks, and suggested next steps.

Column 5: **Today**. Pre-visit questionnaires, solicited patient questions, the last progress note you wrote, and the AI Greek Chorus with suggestions (see **The Chaotic Surge of Intelligence** on page 312).

In my view, a Grok-board should be humane and emphasize the patient's identity and goals, so a physician can connect human-to-human and communicate more effectively. It should also increase the signal-to-noise of the information presented. It should make it easy to grok and then act quickly. It should prioritize the most important next actions and make doing the right thing easy.

Thanks to Gregory Makoul for his groundbreaking work on connecting patient values to physician thinking.[1] It was Dr. Makoul who introduced the idea of helping to humanize the patient by asking them to describe "My joys," "My pressures," and "What I want my doctor to know about me." Their responses were then added to the EHR.

Let's continue to debate the psychology of which displays give the quickest, most useful views of the patient. Let's arm-wrestle, not over sorting of the problem list, but over which display leads to better patient care with less cognitive burden and more joy for physicians, the clinical team, and the patient.

In my wildest dream, I am designing a video gamer's chair that helps you use all your senses. What if your left forearm vibrates to warn you of a medication allergy when you're writing a prescription? What if your right calf feels warm if there is kidney impairment affecting your next decision? Your left calf would feel warm in response to liver impairment. What if your lower back vibrates if you're about to close a chart with unaddressed care gaps. What if we had smell-o-vision to detect the fruity breath of a patient in keto-acidosis?

> **The Lesson Learned**
>
> We are underusing our senses and our pattern-matching skills. Humans grasp visual data much faster than words. Let's build a way for our physicians and teams to grok the patient and then make it easy to do the right thing. We must intentionally build our humanity into this future.

THE FUTURE—All the Lessons Learned

Intro—I Am a Dinosaur

As one of few remaining dinosaurs who remembers the Before Times, I can see what has been lost in the jump from analog to digital. How might we learn from the last tech phase change? How might we handle the pace of change and redesign our teams to include less experienced colleagues and new AI?

Automation Complacency: "That's Probably Right."

We need productive cognitive friction for growth and mastery. We need cognitive ease to reduce administrative burden. How to balance these opposed forces? With one AI assistant, we risk automation complacency. Instead, a Greek Chorus could offer intentionally divergent opinions and help keep us on our toes. Be the conductor, choose the best path.

Automation Clumsiness: "Tech Can Do That for Me."

How can humans "de-skilling" when AI assistants help? The more tasks we delegate, the more our skills erode. Consider simulation training and adaptive automation that scales its assistance up and down to keep human skills sharp.

Mini-Me & Maxi-Me

In the coming year or two, AI agents will dramatically reduce the burden of interacting with the EHR. After that, regulation and licensing and public acceptance, not technology, will be the limiting factors of the coming exponential curve.

Loss of Fault-Tolerance

We must not only design high-functioning High Tide teams but buttress against Low Tide Bad Days. As informaticists, we must use our Psycho 80 skills as well as our tech savvy to anticipate predictable problems and build fault-tolerant teams and systems.

Loss of the Artisan

How might we keep the best qualities of artisanal work and the personal touch, while acknowledging the need for efficiency? I think informaticists will be key members of their design teams, with knowledge of technical possibilities, clinical workflows, and the Psycho 80 skills to build this future.

Loss of Clinical Intuition

We are underusing our senses and our pattern-matching skills. Humans grasp visual data much faster than words. Let's build a way for our physicians and teams to grok the patient and then make it easy to do the right thing. We must intentionally build our humanity into this future.

CONCLUSION

This book shares stories, setbacks, and successes from my career. I share these lessons to help you navigate rough seas during your own journey.

The sea upon which I sail is information technology, and my ship is the electronic health record. My north star is using technology and relationships to improve patient care.

The lessons, however, apply broadly to any field of work, to any ocean of endeavor.

Look back at the PIGlet questions at the beginning of this book (see page 33). Can you answer them for yourself?

(INTRODUCTION) **The Psycho 80**. Working with humans on a big project is surprisingly difficult, yet incredibly rewarding when you succeed. The Psycho 80 principle helps keep your project on track. The technical work is difficult, but the psychosocial interactions are substantially harder and more important.

(INVESTIGATE) **Bridging the gaps**. How well do you listen, really? Do you deeply understand the work, your organization, and your colleagues? Can you master both data and stories? Can you question assumptions, overcome suspicion, and build consensus from chaos?

(INTRIGUE) **Complexity kills**. It is hard to see things clearly. It is hard to say things simply. It is hard to make things easy. Tell better stories. Master your meetings. Build your case, build your microculture, build a high-performance team. A great idea, ignored, helps none.

(INFLUENCE) **Make change happen**. Bring it together. Use your vision, your skills, and your allies. What tactics will you deploy to elevate your teammates, unstick your projects, and build momentum? How can you stay humble, grow humans, and achieve bigger goals?

(INOSCULATE) **Listen, learn, connect**. Medicine and technology are not natural partners, and yet they must work in tandem. How might we combine these fields? As you grow your influence, deploy your skills: empathy, transparency, and unconditional positive regard.

(THRIVE) **Level up**. Grow your core skills: listening, vulnerability, focus, gratitude, storytelling, clarity, conflict resolution. Apply them. Grow. Repeat.

(FUTURE) **The future is uncertain**. Our tools are changing. The change is accelerating. We must adapt and learn to keep up with technology. On the other hand, human emotions, behavior, and teamwork evolve slowly. Psycho 80 skills will be useful for a long time.

APPENDICES

Appendix 1—Sources Cited/Consulted

What Clinical Informatics Is NOT

1. Bhante Bhante Gunaratana, *Mindfulness in Plain English* (Wisdom Publications, 2011)

Stat-LAN and the Chief Complainer

1. Donald Norman, *The Design of Everyday Things: Revised and Expanded Edition* (basic Books, 2013)

Surgery Orders and Complexity: The Cynefin Framework

1. Snowden, David J., and Mary E. Boone. "A Leader's Framework for Decision Making." *Harvard Business Review*, November 1, 2007. https://hbr.org/2007/11/a-leaders-framework-for-decision-making.

What Is a Sacred Bundle, and Why Do You Need One?

1. Peg Neuhauser, *Corporate Legends and Lore: The Power of Storytelling as a Management Tool*, 1993, McGraw-Hill

What TikTok Teaches Us About Communication

1. https://www.tiktok.com/@austinchiangmd
2. https://glaucomflecken.com/

Paternalism and the Patient Portal

1. Edward Krupat, Susan L. Rosenkranz, Carter M. Yeager, Karen Barnard, Samuel M. Putnam, and Thomas S. Inui. "The Practice Orientations of Physicians and Patients: The Effect of Doctor–Patient Congruence on Satisfaction." Patient Education and Counseling 39, no. 1 (January 2000): 49–59. https://doi.org/10.1016/s0738-3991(99)00090-7.

The Superpower That Is the One-Pager

1. Lewis, Michael. "How Two Trailblazing Psychologists Turned the

World of Decision Science Upside Down." *Vanity Fair*, November 14, 2016. https://www.vanityfair.com/news/2016/11/decision-science-daniel-kahneman-amos-tversky.

2. Edward R. Tufte, *The Cognitive Style of PowerPoint: Pitching Out Corrupts Within* (Graphics Press, 2003), 7

The DOs and DON'Ts of an Effective Presentation

1. Arthur C. Clarke, *Electronic Tutors, Omni Magazine*, June 1980

"Put Those People Back Where You Found Them!"

1. https://en.wikipedia.org/wiki/Net_promoter_score
2. Fred Reichheld, *The Ultimate Question 2.0: How Net Promoter Companies Thrive in a Customer-Driven World* (Harvard Business Review Press 2011)
3. Amber Sieja, Katie Markley, Jonathan Pell, Christine Gonzalez, Brian Redig, Patrick Kneeland, and Chen-Tan Lin. "Optimization Sprints: Improving Clinician Satisfaction and Teamwork by Rapidly Reducing Electronic Health Record Burden." Mayo Clinic Proceedings 94, no. 5 (May 2019): 793–802. https://doi.org/10.1016/j.mayocp.2018.08.036.

Sprint: Setbacks and Evolution

1. Amber Sieja, Melanie D Whittington, Vanessa Paul Patterson, Katie Markley, Heather Holmstrom, Stephen Rotholz, Christine Gonzalez, Michael Scott Carpenter, Chen-Tan Lin, "The influence of a Sprint optimization and training intervention on time spent in the electronic health record (EHR)," *JAMIA Open*, Volume 4, Issue 3, July 2021, ooab073, https://doi.org/10.1093/jamiaopen/ooab073
2. Lin CT, Bookman K, Sieja A, Markley K, Altman RL, Sippel J, Perica K, Reece L, Davis C, Horowitz E, Pisney L, Sottile PD, Kao D, Adrian B, Szkil M, Griffin J, Youngwerth J, Drew B, Pell J. "Clinical informatics accelerates health system adaptation to the COVID-19 pandemic: examples from Colorado," J Am Med Inform Assoc. 2020 Dec 9;27(12):1955-1963. doi: 10.1093/jamia/ocaa171. PMID: 32687152; PMCID: PMC7454679.

3. English EF, Holmstrom H, Kwan BW, Suresh K, Rotholz S, Lin CT, Sieja A. "Virtual Sprint Outpatient Electronic Health Record Training and Optimization Effect on Provider Burnout," Appl Clin Inform. 2022 Jan;13(1):10-18. doi: 10.1055/s-0041-1740482. Epub 2022 Jan 5. PMID: 34986492; PMCID: PMC8731238.

Machiavelli on the Inevitability of Resistance

1. "Getting Rid of Stupid Stuff: The Original Launch | Workflow and Process | AMA Steps Forward | Ama Ed Hub." AMA Ed Hub. Accessed May 14, 2026. https://edhub.ama-assn.org/steps-forward/module/2758834.

Machiavelli on The Burning Platform

1. Amber Sieja, Katie Markley, Jonathan Pell, Christine Johnston, Robert Peskind, Chen-Tan Lin, "Successful Implementation of APSO Notes Across a Major Health System," Population Health, Equity & Outcomes Volume 5 Issue 1 , March 2017

The Ferret and the Flywheel

1. Collins, Jim; Hansen, Morten T. *Great by Choice* (Harper Business, 2011)
2. Everett M. Rogers, *Diffusion of Innovations* (The Free Press, 1962)

Don't Be Selfish, Be Useful

1. Richard T. Pascale, *Managing on the Edge: How the Smartest Companies Use Conflict to Stay Ahead* (Simon & Schuster, 1991), 102

Grow Your Team: Square Peg, Round Hole

1. Patterson, Kerry; Grenny, Joseph; McMillan, Ron; Switzler, Al; *Crucial Conversations: Tools for Talking When Stakes Are High* (McGraw Hill, 2011)
2. Fisher, Roger; Ury, William L.; Patton, Bruce; *Getting to Yes: Negotiating Agreement Without Giving In* (Penguin Books, 1991)
3. Stone, Douglas; Heen, Sheila; *Thanks for the Feedback: The Science and Art of Receiving Feedback Well* (Viking, 2014)

4. Collins, Jim, *Good to Great: Why Some Companies Make the Leap... and Others Don't* (Harper Business, 2001)

Grow Your Team: Grow Humans, Not Employees

1. Collins, Jim, *Good to Great: Why Some Companies Make the Leap... and Others Don't* (Harper Business, 2001)
2. Lythcott-Haims, Julie, *How to Raise an Adult* (Holt Paperbacks, 2016)
3. Chris Duffy, host, *How To Be A Better Human*, "How to be an adult – and how to raise one (w/ Julie Lythcott-Haims)" TED, June 5, 2023, https://www.ted.com/podcasts/how-to-be-a-better-human/how-to-be-an-adult-and-raise-one-w-julie-lythcott-haims-transcript

The Information Blocking Rule

1. Reza Ghafoorian, Esq. "Understanding the Information Blocking Rule: Compliance and Exceptions" (presentation, February 24, 2021)
2. Steitz BD, Turer RW, Lin C, et al. "Perspectives of Patients About Immediate Access to Test Results Through an Online Patient Portal," JAMA Netw Open. 2023;6(3):e233572. doi:10.1001/jamanetworkopen.2023.3572
3. California Senate, CA SB1419 | 2021-2022
4. Colorado Senate, SB 26-162, "Sensitive Test Results in Health Care," 75th General Assembly

Fetal Demise Changes an Organization

1. Reza Ghafoorian, Esq. "Understanding the Information Blocking Rule: Compliance and Exceptions" (presentation, February 24, 2021)
2. Rotholz, S; Lin, CT. "'I don't think it should take you three days to tell me my baby is dead.' A case of fetal demise: unintended consequences of immediate release of information," J Am Med Inform Assoc. 2023 Jun 20;30(7):1301-1304. doi: 10.1093/jamia/ocad074.

A Rising Tide in the MMOLC

1. "Information Blocking." Office of Inspector General | Government

Oversight | U.S. Department of Health and Human Services, September 4, 2025. https://oig.hhs.gov/reports/featured/information-blocking/.

OurNotes

1. Walker J, Leveille S, Kriegel G, Lin CT, Liu SK, Payne TH, Harcourt K, Dong Z, Fitzgerald P, Germak M, Markson L, Jackson SL, Shucard H, Elmore JG, Delbanco T, "Patients Contributing to Visit Notes: Mixed Methods Evaluation of OurNotes," J Med Internet Res 2021;23(11):e29951 doi: 10.2196/29951

The Cloud Inside the Triangle

1. John Kotter, *Accelerate: Building Strategic Agility for a Faster-Moving World* (Harvard Business Review Press, 2014)
2. Lin CT, Bookman K, Sieja A, Markley K, Altman RL, Sippel J, Perica K, Reece L, Davis C, Horowitz E, Pisney L, Sottile PD, Kao D, Adrian B, Szkil M, Griffin J, Youngwerth J, Drew B, Pell J. "Clinical informatics accelerates health system adaptation to the COVID-19 pandemic: examples from Colorado," J Am Med Inform Assoc. 2020 Dec 9;27(12):1955-1963. doi: 10.1093/jamia/ocaa171.

Tribes in Medicine and Building Successful Teams

1. Dave Logan, *Tribal Leadership: Leveraging Natural Groups to Build a Thriving Organization* (Harper Business, 2011)

Useful Information Hides Inside a Feeling

1. Fredrik Backman, *A Man Called Ove* (Atria Books, 2015)
2. Daniel Kahneman, *Thinking, Fast and Slow* (Farrar, Straus and Giroux, 2013)

Intro—The Eppendorf Tube and the Boston Winter

1. Daniel Gilbert, *Stumbling on Happiness* (Vintage, 2007)

My Judge Shoots the Second Arrow

1. Gyatso, Tenzin; Tutu, Desmond; Abrams, Douglas Carlton; *The Book of Joy: Lasting Happiness in a Changing World* (Avery, 2016)

The Social Media Power Law

1. Clay Shirky, *Here Comes Everybody: The Power of Organizing Without Organizations* (Penguin Publishing, 2009)

A Ukulele in Healthcare

1. https://ukuleleunderground.com/

Grow a Strength or Patch a Weakness?

1. Atul Gawande, *Better* (Metropolitan Books, 2008)
2. Atul Gawande, "Want to get great at something? Get a coach," TED Talk, Vancouver, BC, April 2022, 16 min., 36 sec., https://www.ted.com/talks/atul_gawande_want_to_get_great_at_something_get_a_coach
3. Bloom, Sahil. "How to Change Your Life in 30 Days: The Curiosity Chronicle." Sahilbloom.com. Accessed May 14, 2026. https://www.sahilbloom.com/newsletter/how-to-change-your-life-in-30-days.

The Startup CEO's Little Black Book

1. Collins, Jim; Hansen, Morten T. *Great by Choice* (Harper Business, 2011)
2. Mueller, Pam A., and Daniel M. Oppenheimer. "The Pen Is Mightier than the Keyboard." Psychological Science 25, no. 6 (April 23, 2014): 1159–68. https://doi.org/10.1177/0956797614524581.

Getting Things Done in an Age of Acceleration

1. David Allen, *Getting Things Done: The Art of Stress-Free Productivity* (Penguin Books, 2003)
2. Cal Newport, *Deep Work: Rules for Focused Success in a Distracted World* (Grand Central Publishing, 2016)
3. Nicholas Carr, *The Shallows: What the Internet is Doing to our Brains* (W. W. Norton & Company, 2011)

4. Collins, Jim, *Good to Great: Why Some Companies Make the Leap... and Others Don't* (Harper Business, 2001)

Get in "The Zone"

1. Mihaly Csikszentmihalyi, *Flow: The Psychology of Optimal Experience* (Harper Perennial Modern Classics, 2008)
2. Ron Chernow, *Alexander Hamilton* (Penguin Books, 2005)

Return to Your Breath

1. https://www.headspace.com/
2. https://store.heartmath.com/
3. https://choosemuse.com/

The Long View: Leading Indicators & the Power of Habit

1. Thomas Sterner, *The Practicing Mind: Bringing Discipline and Focus into Your Life* (New World Library, 2012)
2. Charles Duhigg, *The Power of Habit: Why We Do What We Do in Life and Business* (Random House, 2014)

The Medium View: Accountability & Tuesdays to Write

1. Lowenstein SR. Tuesdays to write... A guide to time management in academic emergency medicine. Acad Emerg Med. 2009 Feb;16(2):165-7. doi: 10.1111/j.1553-2712.2008.00337.x https://medschool.cuanschutz.edu/docs/librariesprovider173/default-document-library/tuesday-stowritearticle.pdf
2. Thomas Sterner, *The Practicing Mind: Bringing Discipline and Focus into Your Life* (New World Library, 2012)

The Short View: Pomodoro Timers & Email Tricks

1. Cirillo, Francesco. The Pomodoro® Technique. Accessed March 31, 2023. https://web.archive.org/web/20230331051358/https:/francesco-cirillo.com/products/the-pomodoro-technique.
2. Barbara Oakley, *A Mind for Numbers: How to Excel at Math and Science*

(Even If You Flunked Algebra) (Tarcher, 2014)
3. https://focuskeeper.co/

Scrolling Alone Vs. Laughing in Groups

1. "The Stanford Model of Occupational Well-BeingTM." The Stanford Model of Occupational Well-Being. Accessed May 15, 2026. https://wellmd.stanford.edu/wellbeing-toolkit/model.html.
2. Robert D. Putnam, *Bowling Alone: The Collapse and Revival of American Community* (Touchstone Books, 2001)
3. "Social Media Bots Infographic," Cisa.gov. Accessed May 15, 2026. https://www.cisa.gov/sites/default/files/publications/social_media_bots_infographic_set_508.pdf.
4. Vivek Murthy, *Together: The Healing Power of Human Connection in a Sometimes Lonely World* (Harper, 2020)
5. Lemann, Nicholas, "Kicking in Groups." The Atlantic, April 1996. https://www.theatlantic.com/magazine/archive/1996/04/kicking-in-groups/376562/.
6. "What is a Balint Group?" Accessed May 15, 2026. https://www.americanbalintsociety.org/what_is_a_balint_group.php.
7. "Wellness in the CU Anschutz Department of Medicine." Accessed May 15, 2026. https://medschool.cuanschutz.edu/medicine/faculty-and-staff-affairs/welldom#ac-wellness-at-the-table-2https://americanbalintsociety.org/.

Write like Hell

1. Czernik, Zuzanna, and Lin CT. "Time at the Bedside (Computing)." JAMA 315, no. 22 (June 14, 2016): 2399. https://doi.org/10.1001/jama.2016.1722.

Intro—I Am a Dinosaur

1. Norman, Jeremy. "'Adventure,' or 'Colossal Cave Adventure,' Is the First Computer Text Adventure Game." History of Information. Accessed May 15, 2026. https://www.historyofinformation.com/detail.

php?id=2020.

The Chaotic Surge of Intelligence

1. Nicholas Carr, *The Glass Cage: Automation and Us* (W. W. Norton & Company, 2014)

Automation Complacency: "That's Probably Right."

1. Nicholas Carr, *The Glass Cage: Automation and Us* (W. W. Norton & Company, 2014)
2. Barbara Oakley, *A Mind for Numbers: How to Excel at Math and Science (Even If You Flunked Algebra)* (Tarcher, 2014)

Mini-Me & Maxi-Me

1. Dubner, Stephen J. "Our Daily Bleg: Did I.B.M. Really See a World Market 'for about Five Computers'? - Freakonomics." https://freakonomics.com/, 2008. https://freakonomics.com/2008/04/our-daily-bleg-did-ibm-really-see-a-world-market-for-about-five-computers/.
2. Kendall, Graham "Your Mobile Phone vs. Apollo 11's Guidance Computer." Real Clear Science. Accessed May 16, 2026. https://www.realclearscience.com/articles/2019/07/02/your_mobile_phone_vs_apollo_11s_guidance_computer_111026.html.

Loss of Clinical Intuition

1. Makoul, Gregory. "Makoul." Wisdomics.org. Accessed May 15, 2026. https://www.wisdomics.org/makoul.

Appendix 2—Web Links

CT Lin's full repertoire of ukulele parodies can be heard on his YouTube channel https://youtube.com/@ctlin1.

See CT Lin's TikTok channel in all its glory at:
https://www.tiktok.com/@ctlin99

The short video on Epic Secure Chat created by CT Lin's team can be seen here:
https://youtu.be/2_2NGmfdRWc

CT's series of blogs on PowerPoint slides:
Part 1 of 4 – https://ctlin.blog/2025/06/11/how-to-give-a-talk-advanced-ct-lins-version-2025-part-1-of-4/
Part 2 of 4 – https://ctlin.blog/2025/06/18/how-to-give-a-talk-advanced-ct-lins-version-2025-part-2-of-4/
Part 3 of 4 – https://ctlin.blog/2025/06/25/how-to-give-a-talk-advanced-by-ct-lin-part-3-of-4-presenting-style/
Part 4 of 4 – https://ctlin.blog/2025/07/02/how-to-give-a-talk-advanced-ct-lins-version-2025-part-4-of-4/

Appendix 3—One-Pagers

Downloadable versions are available at https://ctlin.blog/2022/09/21/how-to-write-an-effective-1-pager-pig-let-training-series/

uchealth

UCHealth Epic
Physician Informatics

OurNotes proposal for system-wide Primary Care implementation

Executive Summary: ***OurNotes*** is a proposal for an MHC Questionnaire to be sent to patients 2 days prior to their appointment (part of e-Check In) to ask: **How have you been since your last visit?** (new symptoms, health worries, medication changes) and, **What 3 questions do you have for your upcoming visit with your provider?**

Background: From July 2018 to July 2019, 2 clinics (**WISH**, and **Lowry** Internal Medicine) participated in a year-long study. Over 2000 patients and about 40 providers participated. All patients with an RPV (return patient appointment) were included. Annual, preventive, procedure, acute visits were excluded. Patient responses automatically inserted into the standard APSO or APSO MA Support note. Patients are warned that notes are NOT SEEN until they arrive at the visit. Response rate to MHC-Q's is 17-20%, similar to e-Check in rates system wide. Providers and patients were surveyed at end of study. Survey tables are attached. In brief:

PROVIDERS

65% of providers usually or always used patient responses.
100% felt NO CHANGE in time spent writing notes
90%: patient interim histories were SOMEWHAT to VERY useful
95%: patient questions/priorities were SOMEWHAT to VERY useful
40% felt patient's submission was VERY important

PATIENTS:

73% Agree: Questions help prepare for the visit
87% Agree: Sending update of my health before visit is a good idea
89% Agree: Sending my issues to focus on before visit is a good idea
43% Yes, 32% No effect: Did your answers change conversation between you and your provider

Provider: *I felt like it was helpful because it allowed me to prioritize certain items and make sure the patient was satisfied with their visit. It was also helpful in the case that we were short on time because I was able to use that as a tool to say "well, we don't have time to talk about everything, but at least we covered your main concern.*

Patient: *The very simple nature of the questions and the immediacy just before the visit are very well done. In hindsight this seems like a no-brainer to do for every patient before every visit. If nothing else it prompts the patient to think about their care before showing up at the doctor's office.*

Example (Most responses are much shorter), inserted in HPI

Patient Entered Data: How have you been since your last visit?
I'm a healthy 76 year old not on any prescription medications. My current situation is that I travel to NYC frequently to take care of my 12 year old grandson when my son travels for work. My son is a single parent since my daughter-in-law suffers from severe mental illness, lives separately and pre-pandemic had only supervised visits with my grandson. I was last in NYC for 3 weeks last part of February through March 8. My son returned from work travel in France, Morocco, Dubai, Shanghai and Seoul. He returned with respiratory sx which worsened to an array of COVID 19 sx over 2 weeks. Testing wasn't available since he didn't need hospitalization. I overlapped with him for 2 days on his return. On my return to Denver I developed some COVID like sx and self-quarantined but wasn't tested since I didn't get that sick. Because he is going back to work my son was tested for COVID antibodies last week and had a high antibody titre. I am needed again in NYC and am requesting antibody testing to evaluate my risk in returning. Medicare covers this test which is available at National Jewish by appointment with a physician referral. You may
1) contact their Physician Line at 1.800.652.9555, weekdays, 8 a.m. to 5 p.m. Mountain Time or email physicianline@njhealth.org OR
2) Download and FAX the referral form at https://www.nationaljewish.org/NJH/media/Coronavirus/COVID-19-Test-Referral-Form.pdf
What are the top 3 questions for your visit?
Q1. Obtaining covid19 antibody test
I have reviewed and verified the above text

Concerns:

1. **Patients entering data that was NOT SEEN** during the visit (not in MA workflow, patients often asked BRIEF HPI questions or MA unaware of Patient responses (PLAN: insert MHC-Q result into ROOMING Tab)
2. **Potentially lengthening pre-check in time** VS improved gathering of patient agenda
3. **Some clinicians do not use Standard APSO** or APSO MA support: this can be addressed
4. **Response rate** to MHC-Q is low (can be encouraged); does not exclude better MHC-Q in the future.

ASSESSMENT/RECOMMENDATIONS:

1. **We propose expanding to ALL primary care practices** throughout UCHealth including CU Medicine and UCHealth Medical Group and offering to Community Connect practices, potentially to specialty practices.
2. **Correct the technical issue: add MHC-Q responses to Rooming tab**, just above Visit Info so both MA's and providers can view ahead of time, in addition to inserting it automatically into the progress note.

Submitted to Primary Care Steering Committee May, 2020
Respectfully, C. T. Lin, MD

uchealth

Physician/APP Informatics Group

We improve physician and team wellness and effectiveness by building extraordinary relationships and innovative tools.

Sharing Test Results Immediately with Patients: Open Results and the need for Anticipatory Guidance

White paper for Physicians/APPs/Clinical Staff: March 2024, CT Lin & Physician Informatics Group

Executive Summary: The Information Blocking Federal Rule (2021) requires release of test results to patients immediately upon request. Most patients want this. For some, this has caused significant anxiety. **Anticipatory Guidance** at the time of placing orders should be the UCHealth standard. Anticipatory Guidance both satisfies the law AND is the right thing for patients. Here is why and how.

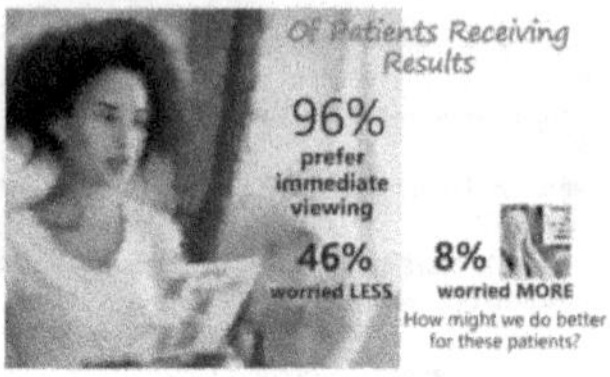

Background: The 21st Century CURES Act makes it illegal for health systems to withhold results from patients by delaying release. UCHealth complied with the law in 2021.

- **CT scan showing cancer**. Dr. Robert Breeze, neurosurgeon, uses anticipatory guidance. At time of ordering the test (CT, MRI, etc), he explains that the patient may see the result before he will. He explains the range of possible results and the actions they will take. Consequently, the timing of release and the next conversation is not time-critical, and patients appreciate the forethought. It is not more work, it is time-shifted upstream.
- **Data**. A four-health-system study* in 2022 (including UCHealth) of 8000 patients receiving test results online indicated that **96% of patients seeing results online (normal or abnormal) WANTED to continue to receive results immediately**. 8% indicated they worried more.
- **Threat, Uncertainty and Lack of Control** are what patients face when seeing new diagnoses and results. From the literature, the concept of **Anticipatory Guidance** addresses these concerns up front and reduces uncertainty.

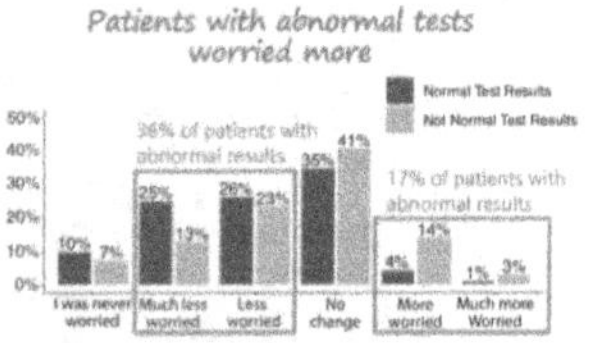

Assessment/Recommendations

For Staff (RNs, MAs): "We release all test results online to you. But, everyone is different. You can choose to look up the result right away or wait for us to contact you. ***Medlineplus.gov*** is an excellent online tool about test results." Also, **you can turn on/off your notifications** in your UCHealth app or UCHealth My Health Connection site (Settings: Notifications, Communications)

For Providers (MD, DO, APP). We recommend universally using **Anticipatory Guidance** at the time of placing orders as follows:

- ***I'm ordering a test****. You might see the result before me.*
- ***You have a choice****: Look immediately or wait to hear from me.*
- *Here is a trusted resource about test results:* ***medlineplus.gov****.*
- ***Here are the best and worst cases****, and how we would contact you.*
- ***What questions do you have****?*

PROPOSAL

Move the discussion upstream to the time of ordering

PRINCIPLES	ANTICIPATORY GUIDANCE
Reduce Threat and Anxiety	Anticipate range of results
Reduce Uncertainty	Anticipate notification process
Improve Sense of Control	Reinforce Choice. Ask for Q

*Steitz, Turer, Lin, et al. 2023, JAMA Network Open 2023

uchealth

EHR Optimization Sprints

Executive Summary, Sep 2024

Summary

The Sprint is a high-performance EHR optimization team. *We engage clinicians & staff w/ empathy & expertise to improve clinical workflows and EHR efficiency so they can regain control of their workload and support their teams.*

Sprint NPS

Net Promoter Score
0-30: good
50: excellent.

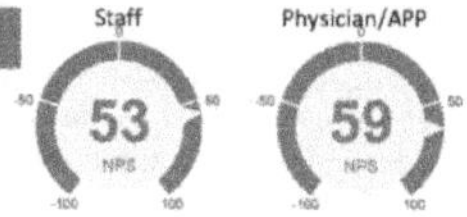

***Provider* Epic Rating** +29
NPS pre-Sprint 23, post: 48

***Staff* Epic Rating** +16
NPS pre-Sprint 32, post: 50

Minutes saved per Provider per Day **16**

Provider/Team Gratitude

RN Navigator: *Sprints are MAGIC. There are so many things they've given me that I had no idea were possible.*

MD: *I feel like it was the best holiday gift you can receive*

MD: *You and your team are INCREDIBLE. I wish you were around all the time on TEAMS.*

Manager: *I feel like Sprint brings so much energy to a clinic. I love them. This is my second Sprint.*

Manager: *All of my staff ... came back raving about the new processes or tidbit they picked up. Thank you to Sprint.*

Manager: *Your staff raised us up to the next level.*

MD: *Can I just tell you that the providers have given such positive feedback about Sprint this year ... You all have truly helped with sooo many things.*

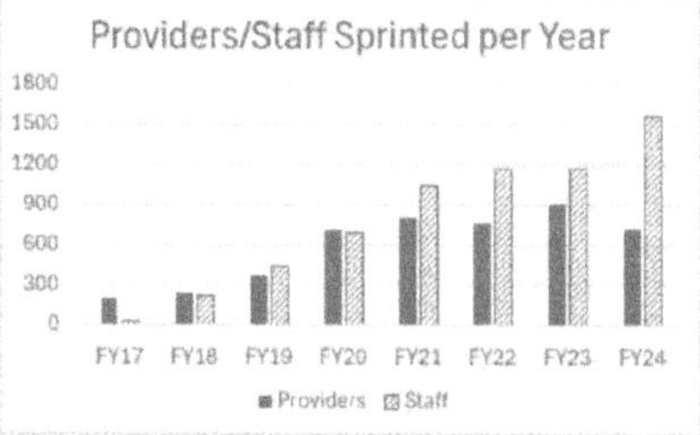

Patient Experience

Our patients tell us:

- Look at our questionnaires (don't waste our time)
- Reconcile our data (dx, meds) (keep updated records)
- Tell us our test results (we are waiting)
- Be inclusive (use the right name and pronouns)

Sprint consistently trains these patient-centered principles.

Training sessions well attended

We sprinted 2300 physicians/APP/staff in FY24

Attendance: Providers: 75%, Staff: 81%, Managers: 71%

Number of 1:1 training sessions: 1700, Group sessions: 80

Organizational Value

Clinic Clean Up. Remove outdated EMP, pools, records. Improve flowsheets, synopses, msg routing, chat groups.

Reduce Unnecessary Variation. Order pref lists, APSO, jump links, disappearing help text, chart filters, quick actions

Improve Standard Work. Letters, reports, patient portal improvements, care gap and quality metric tracking

New Tools: *Requested: 1400. Completed: 1000*

Nurse navigators across system: Built new tools and workflow: dashboards, patient portal strategy, tracking, questionnaires.

PCP and Pop health: pharmacy dispense reports, updated risk scores, message efficiencies. Advance Care Planning flow.

Neurology: new smartsets, flowsheets, intake Q, note templates, bill streamlining, ticket scheduling, care gaps

National/International Reputation

FY25 Planning: 21 Sprints per year

Medical Director, Sprint: Amber Sieja

***Project Management**: Christine Gonzales, Sarah Hutchings. **Physician Informaticists**: Eden English, Heather Holmstrom, Daniel Goodiell, Laura Macke, Rachel Rodriguez, Kate McCaffrey. **Nurse informaticists**: Tory Maklosky, Mary Campbell. **Trainer/Coaches**: Ali Herrera (lead), Beth Rice (lead), Sara Bottger, Joslyn Blatter, Lisa Jesperson, Lesley Evans, Hilary Scholl, Demetria Garduno. **Analysts**: Jill Purdy (manager), Nicky Norton, Jeff Langerak, Kent Spies, Kelsey Brawley, Donna Abdeni, Jemila Kerr, Carlos Chavez*

Surgical Case Request Order (CRO)

Project Summary, November 2024 uchealth

Physician/APP Informatics Group

We improve physician and team wellness and effectiveness by building extraordinary relationships and innovative tools

Summary

Previously, case request orders (CRO) used sticky notes, paper and faxes. This highly variable process caused patient frustration, wrong-site surgery, cancelled procedures, and financial losses due to inaccurate CPT codes. Over 12 months, surgeon, operations, IT and informatics leaders developed consensus on standard process and deployed the Epic electronic CRO. With high CRO adoption by surgery teams across UCHealth, CPT accuracy, OR finances and clinical outcomes have improved.

Quotes from Surgeons/Team

Surgeon: *I found it easy. Saved my favorites. 1-click, done!*

Clinic scheduler: *This is much easier than faxing and email.*

Surgeon: *I have noticed fewer last-minute procedure cancellations due to CPT mismatches.*

Surgeon: *I wasn't looking forward to this, but it wasn't a big deal at all.*

OR scheduler: *the depot is a quick way to glance at planned cases. No faxes are lost anymore. We are not going back.*

Clinic manager: *The dashboards are a big help. We can pinpoint if we have a problem and have data to solve it.*

Prior Organizational Challenges

Patient Safety. 1 wrong procedure, 57 errors that reached pts, 145 near-misses caught before reaching patients (2021)

Incorrect equipment. Linked to surgical delay in 14% of cases

Patient Experience. Over 50% of inaccurate patient estimates are because CPT was incorrect. Required for all surgeries.

Denials. Over $500k in surgery procedures are denied every month due to lack of authorization, due to CPT inaccuracy

Incorrect CPT linked to inefficient block utilization

Case Request Order screen

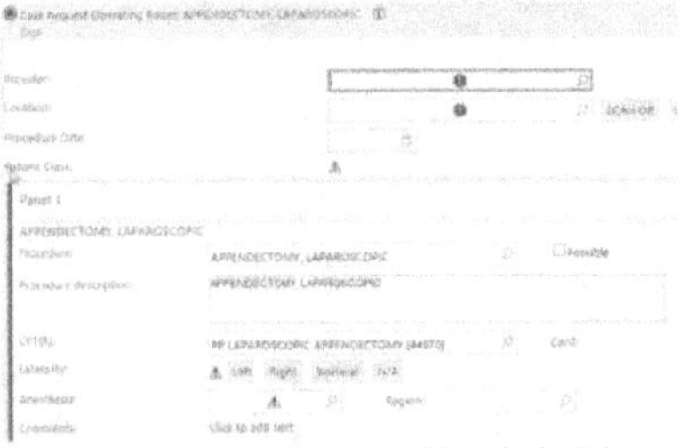

Data entry shared by surgeons, APP's and schedulers

CRO adoption

0 ➜ 89%

Metro: 3000 cases 89%

NoCO: 433 cases 80%

SoCO: 1002 cases 94%

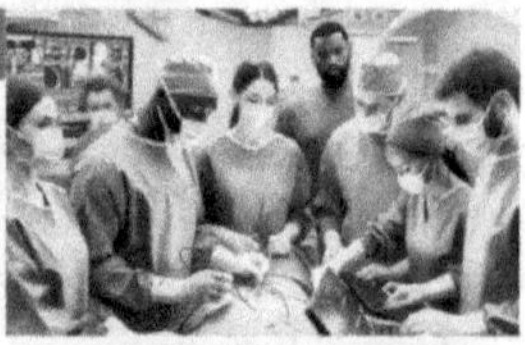

Gen-AI illustration of CRO helping surgery teams

Schedule and Impact of CRO project

Go-lives completed in regional waves:

Metro 11/2023, 1/2024; **South** 3/2024, **North** 4/2024.

270 Surgery clinics and **800** surgeons using CRO

200 Schedulers using CRO

75 IT and other team members supported CRO project

CRO Mismatches Improving

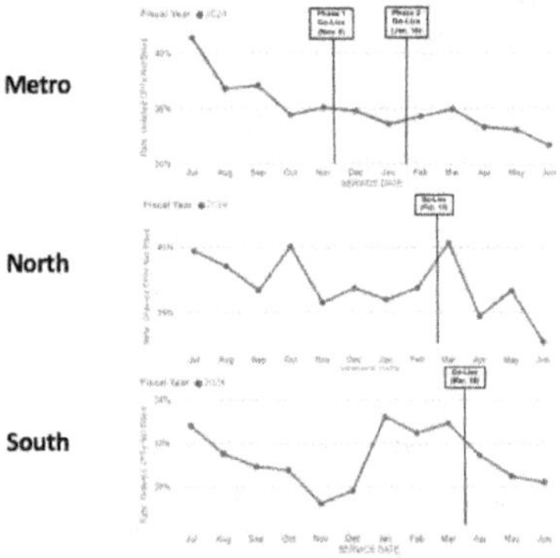

Next Steps

CRO dashboard now available to peri-op leaders for CPT and other process improvement opportunities

Update Financial, Outcome data related to CPT accuracy

Begin e-Consent project for surgeries/procedures

Establish new **IT Perioperative OpTime Steering** group

CRO Team: Raaj Talauliker, Michelle Lieven, Mitch Billimack, Melissa Greer, David Corry MD, Jeniann Yi MD, CT Lin MD

Appendix 4—Additional Reading

Mark Hurst, *Bit Literacy: Productivity in the Age of Information and E-mail Overload* (Good Experience, 2017)

James Clear, *Atomic Habits: An Easy & Proven Way to Build Good Habits & Break Bad Ones* (Avery, 2018)

Gary Keller, *The ONE Thing: The Surprisingly Simple Truth About Extraordinary Results* (Bard Press, 2013)

Nir Eyal, *Indistractible: How to Control Your Attention and Choose Your Life* (BenBella Books, 2019)

About the Author

CT Lin MD, FACP, FAMIA, is a secret introvert who would rather curl up with a good book or compose a clever EHR-themed ukulele parody. He has spent a 30-year career caring for patients and colleagues as an internal medicine physician, informaticist, and Chief Medical Information Officer. The opinions (and inevitable mistakes) in this book are his alone and do not reflect the official positions of either University of Colorado or UCHealth.

Index

www.ingramcontent.com/pod-product-compliance
Lightning Source LLC
LaVergne TN
LVHW010711061126
851857LV00014B/1138